FIELD MANUAL
No. 5–25

HEADQUARTERS
DEPARTMENT OF THE ARMY
WASHINGTON, D.C. *26 May 1967*

EXPLOSIVES AND DEMOLITIONS

*This manual supersedes FM 5-25, 29 October 1963.

ACO 7258A

CHAPTER 1

DEMOLITION MATERIALS

Section I. INTRODUCTION

1. Purpose and Scope

a. This manual is a guide in the use of explosives in the destruction of military obstacles, and in certain construction projects. The material includes information on—

(1) Types, characteristics, and uses of explosives and auxiliary equipment.

(2) Preparation, placement, and firing of charges.

(3) Charge calculation formulas.

(4) Deliberate and hasty demolition methods for use in the forward zone.

(5) Safety precautions.

(6) Handling, transportation, and storage of explosives.

b. The contents of this manual are applicable to nuclear and nonnuclear warfare.

2. Comments

Users of this manual are encouraged to submit comments or recommendations for improvement. Commends should be referenced to the specific page, paragraph, and line of text. The reasons should be given for each to insure proper understanding and evaluation. Comments should be forwarded directly to the Commandant, U. S. Army Engineer School, Fort Belvoir, Virginia, 22060.

3. Military Demolitions

Military demolitions are the destruction by fire, water, explosive, and mechanical or other means of areas, structures, facilities, or materials to accomplish a military objective. They have offensive and defensive uses: for example, the removal of enemy barriers to facilitate the advance and the construction of friendly barriers to delay or restrict enemy movement.

Section II. MILITARY EXPLOSIVES AND SPECIAL CHARGES

4. Definitions

a. Explosives. Explosives are substances that, through chemical reaction, violently change and release pressure and heat equally in all directions. Explosives are classified as low or high according to the *detonating velocity* or speed (in feet per second) at which this change takes place and other pertinent characteristics.

b. Low Explosives. Low explosives *deflagrate* or change from a solid to a gaseous state relatively slowly over a sustained period (up to 400 meters or 1312 feet per second). This characteristic makes low explosives ideal where pushing or shoving effect is required. Examples are smokeless and black powders.

c. High Explosives. The change in this type explosive to a gaseous state—detonation—occurs almost instantaneously (from 1,000 meters per second (3,280 feet) to 8,500 meters per second (27,888 feet), producing a shattering effect upon the target. High explosives are used where this shattering effect is required—in certain demolition charges and in charges in mines, shells, and bombs.

d. Relative Effectiveness Factor. Explosives vary not only in detonating rate or velocity (feet per second), but also in other characteristics, such as density and heat production, that determine their effectiveness. They vary so much that the amount of explosive used is computed according to a relative effectiveness factor, based on the effectiveness of all high explosives in relation to that of TNT. For example, TNT, with a detonating velocity of 23,000 feet per second, has a relative effectiveness factor of 1, while tetrytol, with the same velocity, has a higher relative effectiveness factor of 1.20 (table VIII).

5. Characteristics of Military Explosives

Explosives used in military operations have certain properties or characteristics essential to their function. These are—

a. Relative insensitivity to shock or friction.

b. Detonating velocity, adequate for the purpose.

c. High power per unit of weight.

d. High density (weight per unit of volume).

e. Stability adequate to retain usefulness for a reasonable time when stored in any climate at temperatures between −80° F and +165° F.

f. Positive detonation by easily prepared primers.

g. Suitability for use under water.

h. Convenient size and shape for packaging, storage, distribution, and handling by troops.

i. Capability of functioning over a wide range of temperatures.

6. Selection of Explosives

The explosives for a particular purpose generally are selected on the basis of velocity of detonation. For example, an explosive having a high detonating velocity generally is used for cutting and breaching; that of a lower velocity, for cratering, ditching, and quarrying. The types of explosives commonly used are described below.

7. TNT (Trinitrotoluene)

a. Characteristics (fig. 1).

Case	Color	Size	Detonating velocity
Cardboard with metal ends; threaded cap well.	¼ lb—OD ½ lb—yellow or OD 1 lb—OD	¼ lb—D 1½ in, L 3½ in; ½ lb—3¾ x 1¾ x 1¾ in; 1 lb—7 x 1¾ x 1¾ in.	23,000 fps

Relative effectiveness	Water resistance	Low temperature effects	Packaging
1.0	Excellent (does not readily absorb water).	Less sensitive to shock.	¼ lb—200 blocks in wooden box; ½ lb—100 blocks in wooden box; 1 lb—50 or 56 blocks in wooden box.

b. Use. TNT is used in cutting and breaching and as a main or booster charge for general demolition purposes in combat areas. To form a charge to fit special targets, it is removed from the package and melted in a double boiler. Then it must be immediately cast in the shape needed, because TNT, when melted, becomes fluid and hardens quickly.

Caution: Only those who are well-informed on the characteristics and reaction of molten TNT should attempt this.
If allowed to boil or crystallize, it becomes supersensitive and detonates at a small amount of shock or exposure to flame.

c. Detonation. TNT may be detonated by military electric and nonelectric blasting caps.

8. Tetrytol

a. M1 Chain Demolition Block.

(1) *Characteristics* (fig. 2).

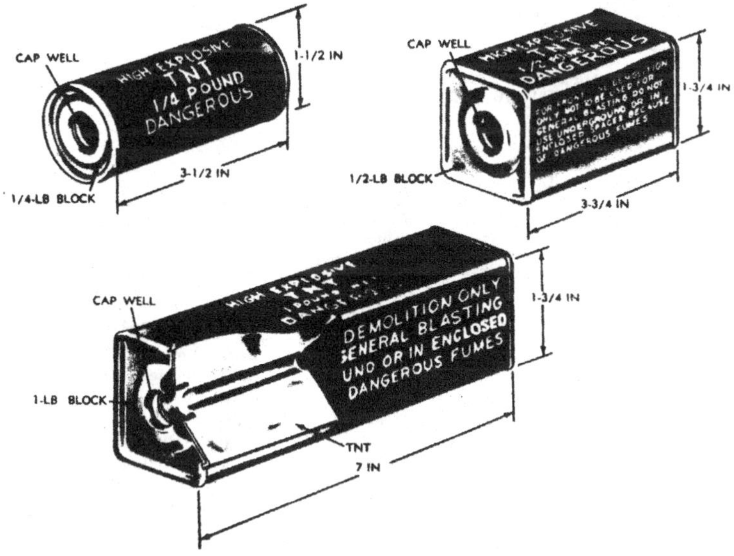

Figure 1. TNT blocks.

Case	Color	Size	Weight	Detonating velocity	Relative effectiveness
Asphalt—impregnated paper wrapper.	**OD**	**11 x 2 x 2 in**	**2⅛ lb**	**23,000 fps**	**1.20**

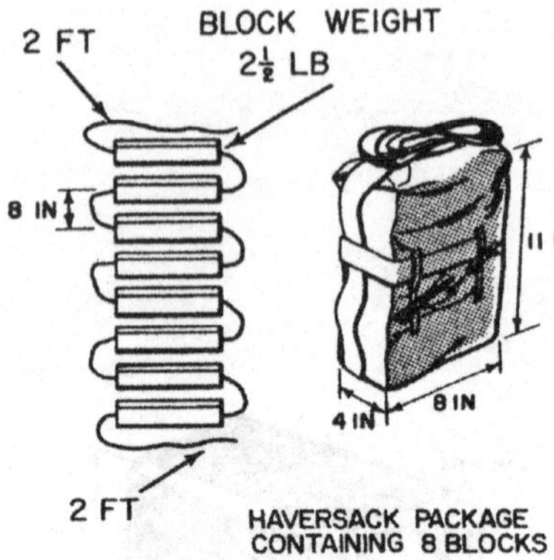

Figure 2. M1 chain demolition block.

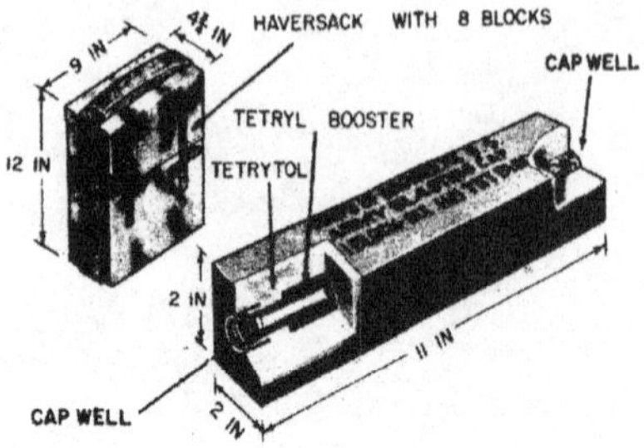

Figure 3. M2 demolition block.

Water resistance	Low temperature effects	Booster	Packaging
Excellent (only slightly soluble).	Less sensitive to shock and slight decrease in strength. Cut ends of detonating cord should be waterproofed to prevent penetration by moisture.	Tetryl pellet cast into block near each end.	Blocks are cast 8 in apart on detonating cord, with 24 in left free at each end of chain (total 192 in). One chain (8 blocks) packed in OD haversack, 11 x 8 x 4 in weighing 20.2 lb and two chains in haversacks in a wooden box.

(2) *Use.* The M1 chain demolition block may be used as an alternate to TNT. The complete chain, or any part of the chain, may be laid out in a line, wrapped around a target, or used in the haversack as it is packed. The entire chain will detonate, even though the blocks may not be in contact with each other. If less than eight blocks are needed, the required number is cut from the chain. Tetrytol is now being eliminated. When present stocks are exhausted, no more will be procured.

(3) *Detonation.* Tetrytol is detonated by means of the military electric or nonelectric blasting cap. The explosive end of the cap should extend toward the charge.

b. M2 Demolition Block.

(1) *Characteristics* (fig. 3).

Case	Color	Size	Weight	Detonating velocity
Asphalt-impregnated paper wrapper. Has threaded cap well.	OD	11 x 2 x 2 in.	2½ lb.	23,000 fps

Relative effectiveness	Water resistance	Low temperature effects
1.20	Excellent (only slightly soluble).	Slight decrease in strength and less sensitive to shock. Requires 6 turns of detonating cord for positive detonation; will explode or ignite under 50-calibre incendiary machine gun fire at subzero temperatures.

Booster	Packaging
Tetryl pellet cast in block surrounds cap well.	Eight blocks packed in a haversack, weighing approximately 22 lb, and two haversacks in a wooden box.

in the same manner as the M1 block. Tetrytol, however, is now being eliminated. No more will be issued after present stocks are exhausted.

(3) *Detonation.* The M2 demolition block may be detonated by the military electric or nonelectric blasting cap.

9. Composition C3 (M3 or M5 Demolition Block)

a. Characteristics (fig. 4).

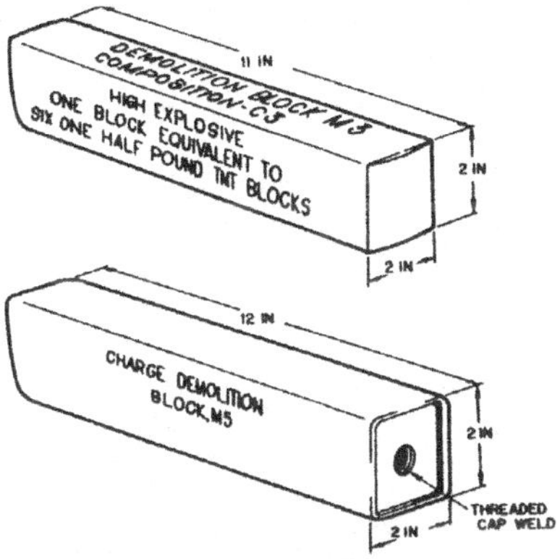

Figure 4. M3 and M5 demolition blocks.

Case	Color	Explosive color	Size
M3—Cardboard wrapper perforated for easy opening; M5—plastic container with threaded cap well.	M3—OD M5—clear plastic	Yellow odorous.	M3–11 x 2 x 2 in M5–12 x 2 x 2 in

Weight	Detonating velocity	Relative effectiveness	Water resistance
M3—2¼ lb M5—2½ lb.	25,018 fps	1.34	Good, but must be in container to prevent erosion.

Low temperature effects	Booster	Packaging	Remarks
When chilled, color changes to red; below −20°F becomes stiff and brittle; plasticity restored by heating. Velocity reduced at −20°F but still of high order.	None	M3–8 blocks packed in haversack and two haversacks in wooden box. M5—1 charge in polyethylene bag, 24 bags in wooden box.	More sensitive than TNT to initiation by impact; and odorous.

b. Use. Because of its plasticity and high detonation velocity, composition C3 is ideally suited to cutting steel structural members. It may be easily molded in close contact to irregularly shaped objects and is an excellent underwater charge if enclosed in a container to prevent erosion.

c. Detonation. Composition C3 may be detonated by the military electric or nonelectric blasting cap.

10. Composition C4

a. M5A1 Demolition Block.

(1) *Characteristics* (fig. 5).

Case	Color	Explosive color	Size
Plastic with threaded recess	Clear plastic	White	11¾ x 2 x 2 in

Weight	Detonating velocity	Relative effectiveness	Water resistance
2½ lb	26,379 fps	1.34	Excellent, if enclosed in original or improvised container to prevent erosion by stream currents.

Low temperature effects	Booster	Packaging	Remarks
Remains like putty at −70° to +170°F. Below −70°, it becomes hard and brittle.	None	One charge packed in polyethylene bag and 24 bags in wooden box.	C4 is more powerful than TNT, without the odor of C3. It is now classified standard B, to be replaced by the M112 demolition charge.

(2) *Use.* Because of its high detonation velocity and its plasticity, Composition C4 is well suited for cutting steel and timber and breaching concrete.

(3) *Detonation.* Composition C4 may be detonated by a military electric or nonelectric blasting cap.

b. M112 Demolition Charge.

(1) *Characteristics* (fig. 5).

Case	Color	Explosive color	Size	Weight	Detonating velocity	Relative effectiveness
Plastic wrapper	White	White	11 x 2 x 1 in	1¼ lb	26,379 fps	1.34

Water resistance	Low temperature effects	Packaging	Remarks
Excellent if inclosed in original or improvised container to prevent erosion by stream currents.	Remains like putty at −70°F to +170°F. Below −70° becomes hard and brittle.	30 blocks per box 14 x 11⅛ x 8 19/32 in; total weight 48 lb.	This is the standard C4 charge replacing the M5A1 block. Sixteen blocks will be available in the M37 demolition kit.

(2) *Use.* Because of its high detonating velocity and plasticity, the M112 demolition charge (C4) is used for cutting steel and timber and breaching concrete. It has an adhesive compound on one face for attachment to target.

(3) *Detonation.* The M112 demolition charge may be detonated by a military electric or nonelectric blasting cap.

11. M118 Demolition Charge

a. Characteristics (fig. 5).

Case	Color	Explosive color	Size	Weight	Detonating velocity
Mylar container	White	Dark green	Block: 12½ x 3¼ x 1¼ in Sheet: 12 x 3 x ¼ in.	Block: 2 lb Sheet: ½ lb	23,616 fps

Relative effectiveness	Water resistance	Low temperatures effects	Packaging
The relative effectiveness factor has not yet been established. For computing test shots, use 1.00.	Unaffected by submersion.	Retains flexibility at −65°F; does not craze or melt at +160°F.	Four sheets per package and 20 packages per box, with a volume of 1.1 cu ft. Total weight 52 lb.

Remarks. May be cut with a knife and placed in an open fire where it will burn but not explode. Will withstand impact of .30 cal. bullets fired from a distance of 40 ft. Each sheet has an adhesive compound on one face.

b. Use. After the protective cover-strip is pulled off, the sheet of explosive may be quickly pressed against any dry surface at a temperature higher than 32° F. A supplementary adhesive has been developed for colder, wet, or underwater targets. The explosive may be used in bulk or cut to accurate width and uniform thickness. It is particularly suitable for cutting steel and breaching.

c. Detonation. The M118 sheet explosive may be detonated by a military electric or nonelectric blasting cap.

12. Composition B

This is a high explosive made of RDX and TNT with a relative effectiveness factor higher than that of TNT (1.35), but is more sensitive. Because of its shattering power and high rate of detonation, Composition B is used as the main charge in certain models of bangalore torpedoes and shaped charges. For further information see table VIII.

13. PETN (Pentaerythritetranitrate)

PETN, the explosive used in detonating cord, is one of the most powerful military explosives, almost equal in force to nitroglycerine and RDX. When used in detonating cord, PETN has a detonation velocity of 24,000 feet per second and is relatively insensitive to friction and shock. For further information see table VIII.

14. Amatol

Amatol is a mixture of ammonium nitrate and TNT with a relative effectiveness of 1.17. Amatol (80/20) may be found in the bangalore torpedo (table VIII).

15. RDX (Cyclonite)

RDX is the base charge in the M6 and M7 electric and nonelectric blasting caps. It is highly sensitive and brisant (great shattering effect) and the most powerful military explosive.

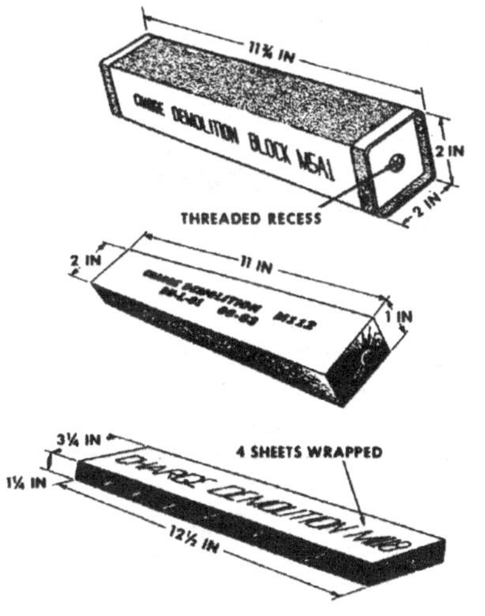

Figure 5. M5A1, M112, and M118 demolition blocks.

16. Pentolite

Pentolite is a combination of PETN and TNT used in the M2A3 shaped charge. Like Composition B it has a high rate of detonation and great shattering power.

17. Ednatol

This is a mixture of halite, or explosive H, and TNT. It has no tendency to combine with metals in the absence of moisture, and has no toxic effects. It is used in shaped charges and high explosive shells.

18. Military Dynamite, M1

a. Characteristics (fig. 6).

Wrapper	Color	Size	Weight	Detonating velocity	Relative effectiveness
Cartridge of waxed paper	Tan	8 x 1¼ in	½ lb	20,000 fps	0.92

Water resistance	Packaging	Remarks
Good	A—50 sticks in waterproofing bag, 2 bags in wooden box B—65 sticks in a carton and 2 cartons in wooden box	Contains no nitroglycerine and thus safer to store, handle, and transport than commercial dynamite. Unaffected by low temperatures.

b. Use. Very satisfactory for construction, quarrying, and many types of demolition work.

c. Detonation. Military dynamite may be detonated by means of a military electric or nonelectric blasting cap, and detonating cord (fig. 79).

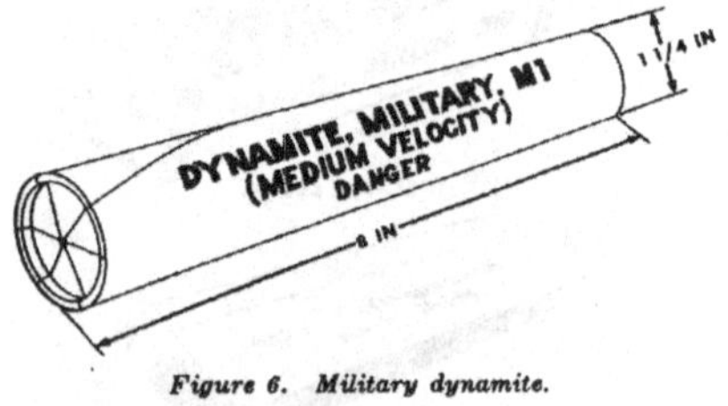

Figure 6. Military dynamite.

19. Commercial Dynamites

a. Introduction. Commercial types of dynamite are straight, ammonia, gelatin, and ammonia-gelatin. Straight dynamites are named according to the percentage of weight of nitroglycerine they contain; for example, 40 percent straight dynamite contains 40 percent nitroglycerine. Ammonia dynamite is different, however, as 40 percent ammonia dynamite indicates that the dynamite is equivalent to 40 percent straight dynamite but not that it contains 40 percent nitroglycerine by weight.

(1) Gelatin dynamite is a plastic dynamite with an explosive base of nitrocotton dissolved in nitroglycerine and is relatively insoluble in water.

(2) Ammonia-gelatin dynamite is a plastic dynamite with an explosive base of nitrocotton dissolved in nitroglycerine with ammonium nitrate added. It is suitable for underwater use.

b. Characteristics.

Wrapper	Color	Size	Weight	Composition
Paraffin-treated paper cartridge.	Tan	8 x 1¼ in	½ lb	Straight—nitroglycerine and nonexplosive filler. Ammonia—ammonium nitrate and nitroglycerine. Gelatin—nitrocotton dissolved in nitroglycerine. Ammonia-gelatin—same as gelatin with ammonium nitrate added.

Detonating velocity		Relative effectiveness	Water resistance
Straight	40 %—15,000 fps 50 %—18,000 fps 60 %—19,000 fps	0.65 0.79 0.83	Good if fired within 24 hours
Ammonia	40 %— 8,900 fps 50 %—11,000 fps 60 %—12,000 fps	0.41 0.46 0.53	Poor
Gelatin	40 %— 7,900 fps 50 %— 8,900 fps 60 %—16,000 fps	0.42 0.47 0.76	Good
Ammonia-gelatin	40 %—16,000 fps 50 %—18,700 fps	--- ---	Excellent

Remarks

Requires careful handling, as flames, sparks, friction, and sharp blows may cause detonation, and special care in storage, as it deteriorates rapidly. It is thus undesirable for military use.

c. Uses. Being sensitive to shock and friction, commercial dynamite is not generally used in forward areas; but it is acceptable in emergencies when other more suitable explosives are lacking. Sixty percent straight dynamite, of less strength than TNT, has a variety of uses; gelatin dynamite is applicable to underwater demolitions and to land clearing, cratering, and quarrying. A gelatin dynamite of low heaving force and a high rate of detonation is used for blasting hard rock.

d. Detonation. Commercial dynamites may be exploded when primed with a commercial No. 6 or larger, blasting cap, a military electric or nonelectric blasting cap, or detonating cord (fig. 79).

e. Low Temperature Effects. The sensitivity of dynamite decreases at diminishing temperatures until the dynamite freezes, after which it becomes extremely sensitive. Gelatin dynamite does not freeze as easily as straight dynamite. When straight dynamite is stored, the nitroglycerine tends to settle out of the sticks; accordingly, straight dynamite cases should be frequently and regularly turned until freezing sets in. Frozen dynamite may be thawed in a kettle as described in *g*, below.

f. Old Dynamite. Old dynamite may be recognized by the oily substance collected on the casing or by stains appearing on the wooden packing case. These are caused by the separation of the nitroglycerine from the porous base. Dynamite in this state, being extremely sensitive, must not be used but destroyed immediately by burning (TM 9–1300–206).

g. Frozen Dynamite. Frozen dynamite is recognized by its hardness and by the appearance of crystals (which are extremely sensitive) in the contents of the stick. Its use is not recommended. It may be destroyed by burning in the same manner as old dynamite. Frozen dynamite, may be used, however, if thawed as follows:

(1) Use a commercial thawing kettle. If this is not available, a 5- and a 10-gallon container may be combined to make a good substitute.

(*a*) Heat water in a separate container to a temperature as high as can be tolerated by the hand.

(*b*) Pour the heated water into the water compartment of the thawing kettle (10 gallon can).

(*c*) Lay the frozen dynamite in the inner compartment (5-gallon container) in a horizontal position, with the bottom sticks supported on strips of wood or other material, so that the air can circulate readily around the sticks.

(*d*) Place the kettle in a barrel or box insulated by hay or some other satisfactory material.

(*e*) Thaw no more than 50 pounds of frozen dynamite in a single lot.

(*f*) Never place the frozen dynamite in the thawing compartment of the kettle before the hot water is poured into the water compartment.

(*g*) Never set the kettle over heat after the dynamite has been placed in it.

(2) Frozen dynamite is completely thawed when it has returned to its original consistency. This can be determined by squeezing the sticks lightly with thumb and forefinger. If no hard spots remain and when unwrapped no crystals are seen, it is thawed and ready for use.

20. Foreign Explosives

a. Types. Explosives used by foreign countries include TNT, picric acid, and guncotton. Picric acid has characteristics like TNT except that it corrodes metals and thus forms extremely sensitive compounds. A picric acid explosive in a rusted or corroded container must not be used; in fact, it should not be handled in any way, except to move it very carefully to a safe disposal area or location for destruction (app C).

b. Uses. Explosives of allied nations and those captured from the enemy may be used to supplement standard supplies. Such explosives, however, should be used only by experienced soldiers and then only according to instructions and directives issued by theater commanders. Captured bombs, propellants, and other devices may be used with U. S. military explosives for larger demolition projects, such as pier, bridge, tunnel, and airfield destruction (app C). Most foreign explosive blocks have cap wells large enough to receive U. S. military blasting caps. These blasting caps, when used to detonate foreign explosives, should be test fired to determine their adequacy before extensive use.

21. Ammonium Nitrate

a. Characteristics (fig. 7).

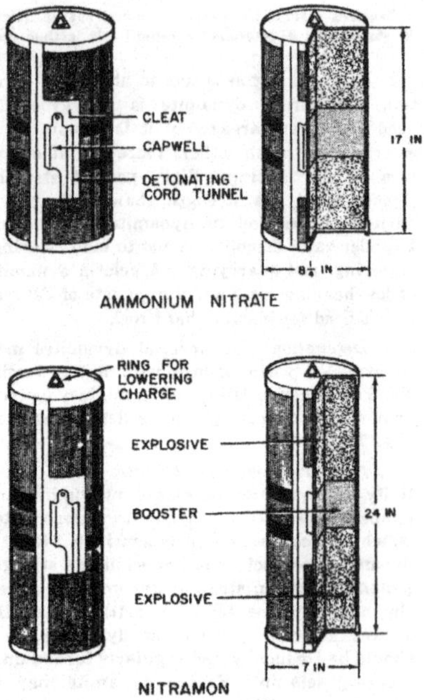

Figure 7. Ammonium nitrate and nitramon cratering charge.

Container	Color	Size	Weight of charge	Detonating velocity	Relative effectiveness
Cylindrical metal	OD	17 x 8¼ in	11,000 fps	40 lb	0.42

Water resistance	Low temperature effects	Booster
Poor. Should not be removed from container in cratering because of moisture absorption.	Slight loss in strength but functions satisfactorily.	TNT surrounding the cap well.

Packaging	Remarks
One charge is packed in wooden box 22½ x 9½ x 9¾ in. Total wt 50.8 lb.	Container has ring on top for handling and lowering into boreholes.

b. Uses. Having a low detonating velocity (11,000 fps) and thus a low shattering power that produces a pushing or heaving effect, ammonium nitrate is used chiefly as a cratering charge. It is also effective in ditching.

c. Detonation. The container has a cap well

and a detonating cord tunnel for priming. A cleat is placed above and to the side of the cap well for attaching electric and nonelectric primers. Frequently a primed one-pound block of TNT is placed on the charge to insure detonation.

22. Nitramon Cratering Charge

a. Characteristics (fig. 7).

Container	Color	Size
Cylindrical metal.	OD	24 x 7 in

Weight of charge	Detonating velocity	Relative effectiveness
40 lb	11,000 fps	0.42

Water resistance	Low temperature effects
Poor. Should not be removed from the metal container in cratering because of moisture absorption.	Slight loss in strength but functions satisfactorily.

Booster	Packaging
TNT surrounding the cap well.	One metal container in wooden box 27½ x 8⅞ x 9¾ in. Total weight 52 lb.

b. Use. Because of the low detonating velocity (11,000 fps) and low shattering power that produces a heaving effect, this charge is very effective in cratering and ditching. The container has a ring on top for general handling and lowering into boreholes.

c. Detonation. The container is fitted with a cap well and a tunnel for priming and a cleat to attach electric and nonelectric primers. A primed 1-pound block of TNT is placed on the charge for positive detonation.

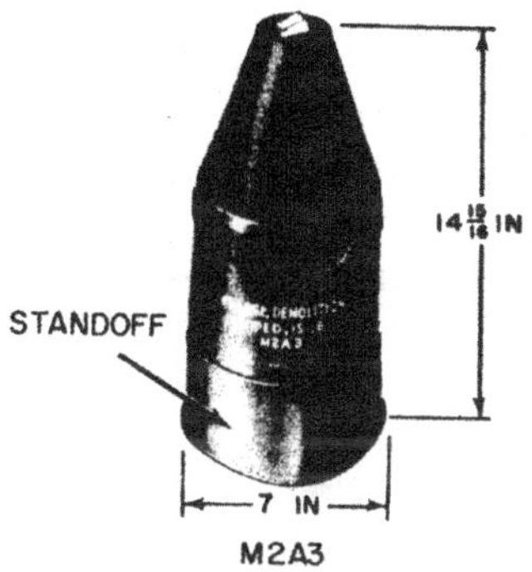

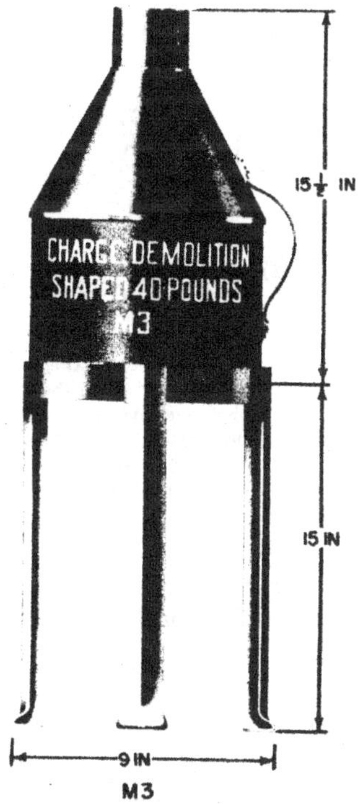

Figure 8. Shaped charges.

23. Shaped Charges

A shaped charge is an explosive charge with its detonating action directed to increase its effectiveness in penetrating steel, armor, and concrete and other masonry (fig. 8). Charges, as issued, are usually cylindrical in shape but may be linear like the charges included in the M157 demolition kit (para 27*c*). Cylindrical shaped charges have a conical top, a conical recess, and a metal or glass liner in the base. The threaded cap well in the top is for priming with military electric or nonelectric blasting caps. Shaped charges, generally, are made from such explosives as Composition B, pentolite, and ednatol.

a. M2A3 and M2A4 Shaped Charges.

(1) *Characteristics* (fig. 8).

Case	Color	Weight of charge	Size	Explosive	Booster
Water resistant fiber	OD	12 lb	14 15/16 x 7 in	50/50 pentolite or composition B.	M2A3: 50/50 pentolite with comp B explosive. M2A4: Comp A3, resistant to small arms fire.

Liner	Low temperature effects	Packing	Remarks
Glass	Satisfactory in arctic climates.	Three charges packed in wooden box 33⅛ x 10⅝ x 9½ in. Total weight 66 lb.	Both models have a cardboard cylindrical standoff fitted to the case.

(2) *Use.* Shaped charges are used primarily to bore holes in earth, metal, masonry, concrete, and paved and unpaved roads. Their effectiveness depends largely on their shape and the material of which they are made, the explosive, and proper placement. The penetrating capabilities in various materials and proper standoff distances are given in table XII.

(3) *Detonation.* The M2A3 and M2A4 shaped charges contain a threaded cap well for detonation by electric and nonelectric blasting caps. Dual priming, however, is extremely difficult because of the configuration of the case and the need for priming at the exact rear-center. They are not effective under water because of the obstruction to the jet.

b. M3 Shaped Charge.

(1) *Characteristics* (fig. 8).

Case	Color	Weight of charge	Size	Explosive	Booster
Metal	OD	30 lb	15½ x 9 in (less standoff).	50/50 pentolite or composition B.	50/50 pentolite with comp B charge.

Liner	Low temperature effects	Packaging	Remarks
Steel or copper	Satisfactory in arctic climates.	One each in wooden box 20½ x 13⅝ x 11⅝ in. Total weight 65 lb.	Provided with metal tripod standoff 15 in high.

(2) *Use.* Shaped charges, primarily, are used to bore holes in earth, metal, masonry, pavement and the like. Effectiveness depends considerably on the shape and material in the cone, the explosive used, and proper placement. The penetrating effects of shaped charges in various materials and relative standoff distances are given in table XII.

(3) *Detonation.* The M3 shaped charge is provided with a threaded cap well for detonation by electric and nonelectric blasting caps. Dual priming is very difficult because of the slope of the case and the need for exact rear-center priming. The M3 shaped charge is not effective under water because of the obstruction to the jet.

c. Special Precautions. In order to achieve the maximum effectiveness of shaped charges—

(1) Center the charge over the target point.

(2) Set the axis of the charge in line with the direction of the hole.

(3) Use the pedestal provided to obtain the proper standoff distance.

(4) Be certain that there is no obstruction in the cavity liner or between the charge and the target.

(5) Be certain that soldiers using shaped charges in the open are at least 900 feet away in defilade under cover, or at least 300 feet away if in a missile-proof shelter, before firing.

24. M1A1 and M1A2 Bangalore Torpedo Kits

a. Characteristics (fig. 9).

Case	Color	Components	Section weight	Section dimensions	Explosive
Metal	OD	10 loading assemblies or sections; 10 connecting sleeves; one nose sleeve.	Approx 13 lb	5 ft x 2 ⅛ in	M1A1—approx 9 lb amatol and TNT booster; M1A2—approx 9 lb comp B and comp A3 booster.

Packaging	Remarks
One kit packed in wooden box 64⅛ x 13⅝ x 7⅛ in. Total weight 176 lb.	Four inches of length at both ends of each section contains a booster.

b. Assembly for Use. All sections have a threaded cap well at each end so that they may be assembled in any order. The connecting sleeves make rigid joints. A nose sleeve is placed on the front of the torpedo to assist in pushing it through entanglements and across the ground. It is also desirable to attach an improvised loading section without explosive on the end to forestall premature detonation by a mine when the torpedo is shoved into place. In the assembly of two or more tubes, the nose sleeve is pressed onto one end of one tube, and the other end is connected to a second tube by a connecting sleeve. A bangalore torpedo or torpedo section may be improvised by the use of a 2-inch diameter pipe with a 24-gage wall thickness with approximately 2 pounds of explosive per foot of length. Successive pipe lengths, however, must be closely connected.

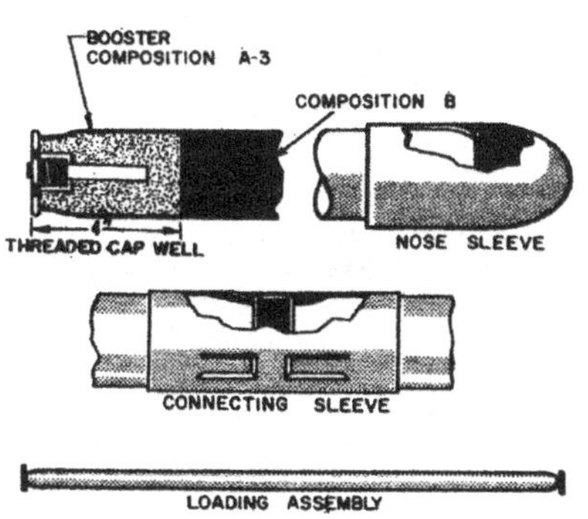

Figure 9. M1A1 bangalore torpedo.

c. Use. The bangalore torpedo clears a path 10 to 15 feet wide through barbed wire entanglements. In minefield breaching, it will explode all antipersonnel mines and most of the antitank mines in a narrow foot path. Many of the mines at the sides however may be shocked into a sensitive state, which makes extreme care necessary in any further mineclearing. Bangalore torpedoes also may be used in bundles as substitute explosive charges in the M3A1 antitank mineclearing projected charge demolition kit (fig. 13).

d. Detonation. The military electric or nonelectric blasting caps will detonate the bangalore torpedo. In obstacle clearance, the bangalore torpedo should be primed after it has been placed. The cap well at the end should be protected with tape or a wooden plug while the torpedo is being pushed into place. Priming is generally done either by means of priming adapter and a military electric or nonelectric blasting cap and time fuse, or by detonating cord with six turns around the 4-inch booster portion of the torpedo.

25. M37 Demolition Charge Assembly

The M37 charge assembly (fig. 10) consists of eight M5A1 demolition blocks, eight demolition block hook assemblies, and two M15 priming assemblies. The demolition blocks are packed in two bags, four blocks per bag, and the assembly placed in an M85 carrying case. The M15 priming assembly is a 5-foot length of detonating cord with two plastic adapters and two RDX boosters attached. The adapters are threaded to fit the standard cap well in the demolition block. The priming assembly has two detonating cord clips for fixing the M37 charge assembly to the main line. The hook assemblies are hooks and pieces of rope for attaching charges to the target.

a. Use. This assembly is applicable to the use of assault demolition teams in the reduction of obstacles. It is very effective against small dragon's teeth approximately 3 feet high and 3 feet wide at the base.

b. Detonation. The M37 demolition charge is detonated by means of the M15 priming assembly and an electric or nonelectric blasting cap or by a ring main attached by means of the detonating cord clips provided.

Figure 10. M37 demolition charge assembly.

c. Packaging. One assembly is packed in an M85 carrying case, and two are packed in a wooden box 17⅛ x 11½ x 12½ inches. The gross package weight is 57 pounds.

26. Rocket-Propelled Train Bangalore Torpedo (Barney Google)

The device consists of 20 sections of bangalore torpedo fitted together by special connecting sleeves to form a 100-foot train (fig. 11). A kit contains the rocket motor, tail assemblies, and couplings for 20 sections. The motor is fitted to the front of the train to provide propulsion. Detonation occurs at the tail assembly by means of a nonelectric blasting cap, pull fuze, and a reel of cable.

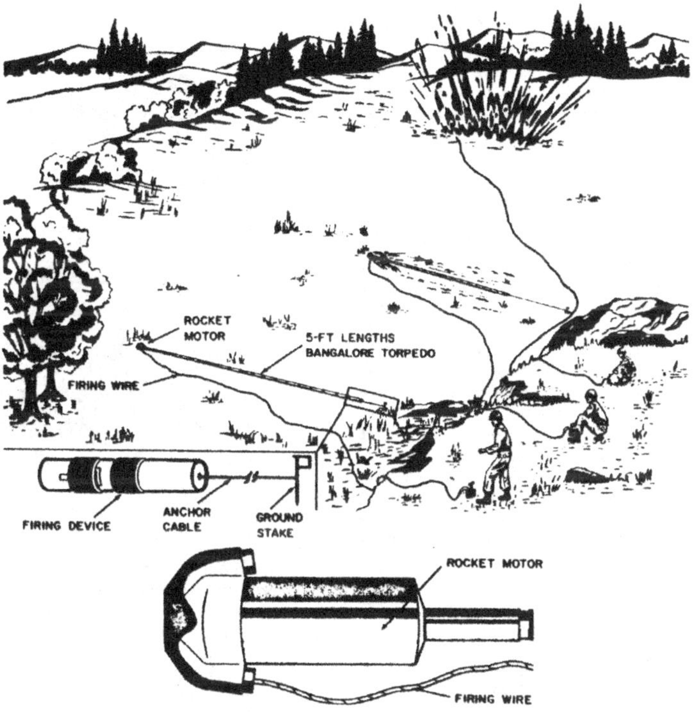

Figure 11. Rocket-propelled bangalore torpedo (Barney Google).

a. Uses. The rocket-propelled bangalore torpedo is used against barbed wire entanglements, antipersonnel mines, and similar small obstacles. The rocket propulsion enables deeper penetration of small obstacles with less exposure of friendly soldiers to enemy observation and fire.

b. Detonation. The assembled torpedo is placed at a spot within range of the target. The 400-foot reel of cable is shortened to the proper length and its free end is anchored firmly. After the safety has been unscrewed from the tail assembly and all soldiers have taken cover, the rocket motor is fired electrically. After the torpedo has traveled a distance equal to the length of the anchored cable, the pull fuze is actuated and the assembly detonated.

27. Projected Charge Demolition Kits

a. M1 and M1E1. These are identical in all respects except for the delay detonators and the time blasting fuze igniters issued with them. The M1E1 has the M60 weatherproof blasting fuse igniter and the 15-second delay M1A2 percussion detonator, while the M1 has the M2 weatherproof blasting fuse igniter and the 15-second delay M1 or M1A1 friction detonator (fig. 12).

(1) *Components.* These are a nylon-covered detonating cable, propulsion unit, launcher, fuse lighter, delay detonator, anchor stake, and carrying case. The explosive item, or detonating cable, is 1 inch in diameter and approximately 170 feet (52 meters) long; it weighs 63 pounds, 46 pounds of which is oil-soaked PETN. The detonating cable is composed of 19 strands of special detonating cord, each containing 100 grains of PETN per foot. This differs from the regular (reinforced) detonating cord, which contains only 50 to 60 grains of PETN per foot. Regular detonating cord cannot be used as a substitute in the kit.

(2) *Use.* This kit is emplaced to project and detonate a cable across a pressure-actuated antipersonnel minefield. Grass, leaves, other light vegetation, and some soil are blown aside in a lane about 8 feet wide. More soil is blasted aside when the ground is moist and soft than when it is dry and hard. Camouflaged antipersonnel mines and those near the surface in the 8 foot lane are usually exposed.

(3) *Detonation.* One soldier fires the complete assembly. First the kit is emplaced; then the fuse lighter on the jet propulsion unit is pulled. The 15-second delay in the propulsion fuse allows the soldier to move the 5-foot distance from the launcher to the anchor stake and pull the fuse lighter safety pin and pull ring on the detonating cable (which also has a 15-second delay) and then take cover at least 100 feet behind the assembly.

(4) *Packaging.* The complete assembly is issued in a OD-colored waterproof aluminum carrying case. Each case is packed in a wooden box 25½ x 18⅜ x 19¼ inches. The gross weight of the kit and box is 142 pounds.

b. Antitank Mine-Clearing Kits. These kits, the M2, M2A1, M3, and M3A1, consist of semi-rigid projected charges and the accessories and tools needed to assemble and attach them to a light or medium tank. They are approximately 14 inches wide, 5 inches high, and 400 feet (121.9 meters) long, weighing approximately 9,000 pounds, including 4,500 pounds of explosive. They are supplied in elliptically-shaped units or elements 5 feet long, containing about 35 pounds of explosive. The M3 consists of an 80/20 amatol charge and a 6-inch crystalline TNT booster at each end. The M3A1 (fig. 13 and table I) consists of Composition B charge with a Composition A-3 booster in each end. Bangalore torpedo explosive elements may be substituted for the standard explosive elements, four for each. Both are initiated by two bullet-impact fuses by fire from the main tank armament or from any 37mm or larger high explosive shell with a super-quick fuse. Both types have a threaded cap well suitable for a standard firing device and an electric or non-electric military blasting cap. This cap well also makes possible the use of the explosive ele-

Figure 12. M1E1 projected charge kit.

ment as a separate expedient charge. In most soils, these charges form a crater about 100 meters (330 feet) long, 5 meters (16 feet) wide, and 2 meters (7 feet) deep.

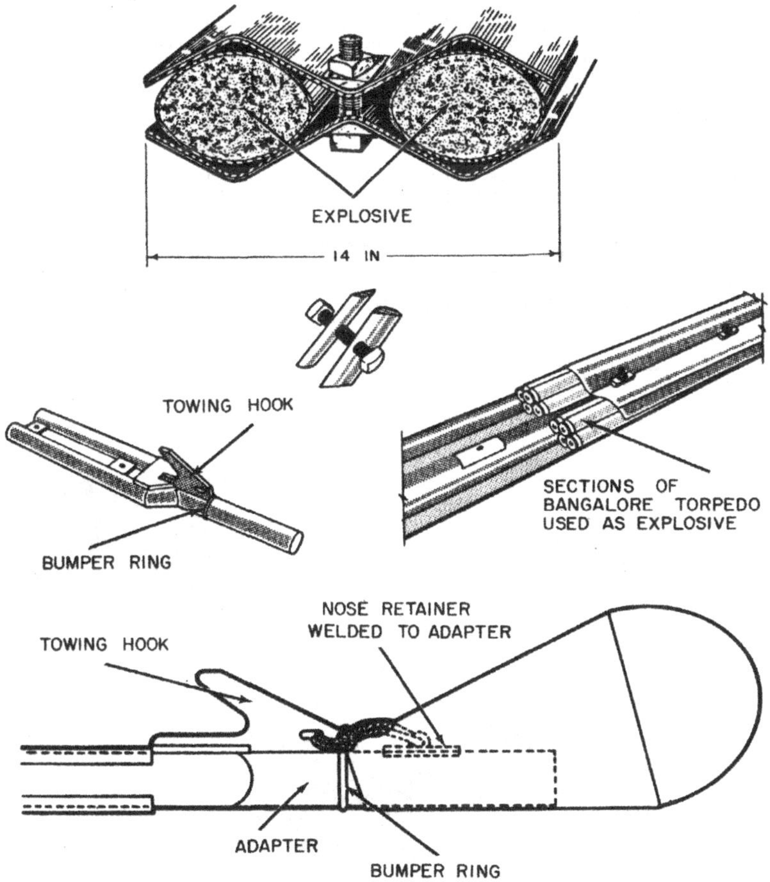

Figure 13. M3A1 projected charge kit.

Table I. Comparison of M2, M2A1, M3, and M3A1 Projected Charge Demolition Kits

	M2	M2A1	M3 and M3A1
Total net weight	12,500 lb	15,000 lb	9000 lb
Corrugated plates	Steel, 53 lb, 164	Steel, 53 lb, 172	Aluminum, 16 lb, 200
Washers	1 per bolt, 2 in long	1 per bolt, 2 in long	2 per bolt, 4 in long
Nose	Steel, two piece, bolted; held to adapter by bolt.	Steel, two piece, bolted; held to adapter by special retainer.	Aluminum, one piece, welded; held to adapter by special retainer.
Tamping bags	Paper	Paper	Cloth or paper
Pushing attachment	Wire rope	Steel chain	Steel chain
Total explosive	3200 lb	4500 lb	4500 lb
Explosive cartridges	4 feet long, 20 lb explosives; steel casing; circular in cross section.	5 feet long, 35 lb explosives; aluminum casing; elliptical in cross section.	5 feet long, 35 lb explosives; aluminum casing; elliptical in cross section.
Fuze and shield	1 of each	2 of each	2 of each
Towing assembly rigging.	Rope on towing yoke raised by hand.	M2 fittings and rigging improved and strengthened; cable on towing yoke raised by winch on periscope fitting.	M2 fittings and rigging improved and strengthened; cable on towing yoke raised by winch on periscope fitting.

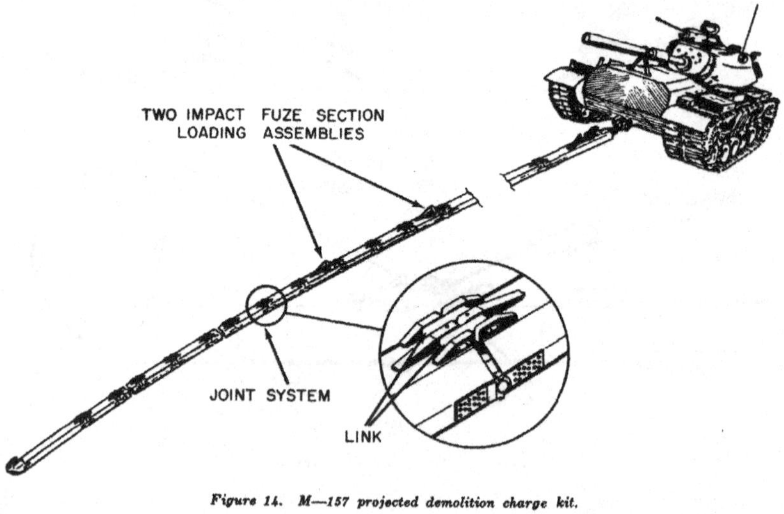

Figure 14. M—157 projected demolition charge kit.

c. M–157.

(1) *Description.* This kit (fig. 14) measures about 12 inches in width, 7 inches in height, and 400 feet (121.9 meters) in length. It consists of 79 sections —1 nose section, 13 body sections, 62 center-loading sections, 2 impact fuse sections, and 1 tail section. Only 64 of the 79 sections contain explosives— the 62 center loading sections and the 2 impact fuse sections. The kit weighs 11,000 pounds including approximately 3,200 pounds of explosive. The explosive is a linear shaped charge, 12 inches wide, 7 inches high, and 5 feet long, containing approximately 45 pounds of Composition B and 5 pounds of Composition C–4. As the insert tubes are welded to the walls of the center loading sections, the explosive elements cannot be used as separate charges or replaced by any substitute item in the field. The linear shaped charge insures a wider, clearer path throughout minefields than many other explosive clearing devices. In most soils this charge forms a crater about 100 meters (330 feet) long, 4 to 5 meters (12 to 16 feet) wide, and 1 to 1½ meters (3 to 5 feet) deep.

(2) *Use.* Projected demolition charges are used chiefly in the deliberate breaching of minefields. They are also effective against bands of log posts, steel rail posts, antitank ditches, and small concrete obstacles. These charges are adequate to break down the sides of an antitank ditch. They will also clear a path through the ditch adequate for tank traffic, if it is unrevetted and 5 feet deep or less and if the charges project beyond the far side of the ditch. They are effective in ditches from 5 to 8 feet deep if the soil is very favorable. The explosive elements of the M3 and M3A1 projected charges may be used as expedient individual charges in the M–157 kit.

(3) *Detonation.* The charges are generally detonated from a tank by means of bullet impact fuse, which has a target plate that bears on the firing pin and is held in place by a shear pin and a safety fork that must be removed before the fuse can be actuated. The fuse is detonated by fire from the main tank armament or from any 37mm or larger high explosive shell with a super-quick fuse. Two fuses are provided to insure that one is visible to the tank gunner at all times.

28. Improvised Charges

Demolition teams operating in the field frequently find targets to which standard methods and charges may not apply and improvisations are required. Frequently the success of the mission depends upon the ingenuity or the team. The package and pole charges are such improvisations. By skillful modifications they may be applied successfully in many situations.

a. Package Charges. Charges prepared in convenient packages of appropriate size and shape are always more readily put in place than other types. Explosives may be packaged in sandbags to make elongated cylindrical charges for boreholes. Blocks of TNT or other explosives may be stacked together and bound with tape or twine or wrapped in canvas, other cloth, or paper. A satchel charge may be improvised by tying or taping explosive blocks to a board with a handle attached. Large charges may consist of an entire case of explosives. Here at least one block or one cartridge is removed from the case, primed, and replaced. A still larger charge may be made by lashing several cases of explosive together. The detonation of a single primer will fire the entire charge. Dual priming systems, however, should be used if possible.

b. Pole Charges. Pole charges are convenient for placement against pill boxes, hard-to-reach bridge stringers, underwater bridge supports, and other locations not easily accessible. Pole charges are usually an assembly of an explosive charge; detonating cord; fuse lighter, time fuse, nonelectric blasting cap; and a pole for placing or propping them in position. Pole charges are usually prepared in the same manner as package charges. Dual priming should be used, if possible.

Section III. DEMOLITION ACCESSORIES

29. Time Blasting Fuse

Time blasting fuse transmits a flame from a match or igniter to a nonelectric blasting cap or other explosive charge, providing a time delay wherein blasters may retire to a safe distance prior to the explosion. There are two types: safety fuse and time fuse M700. These may be used interchangeably.

a. Safety Fuse. Safety fuse is limited standard. It is used in general demolitions. It consists of black powder tightly wrapped with several layers of fiber and waterproofing material and may be any color, orange being the most common (fig. 15). As the burning rate may vary for the same or different rolls from 30 to 45 seconds per foot under different atmospheric and climatic conditions (exposure for over 12 hours to the elements, extreme changes in temperature, and the like), each roll must be tested prior to using in the area where the charge is to be placed. Particular precautions must be taken if used under water, as the rate of burning is increased significantly. Accordingly, each roll should be tested under water prior to preparation of the charge. In arctic temperatures, the outside covering becomes brittle and cracks easily.

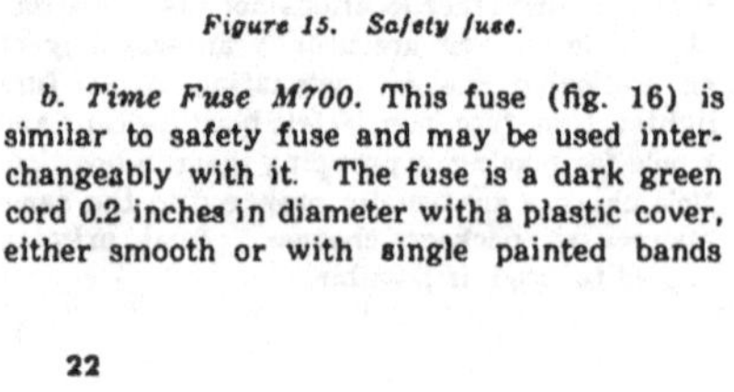

Figure 15. Safety fuse.

b. Time Fuse M700. This fuse (fig. 16) is similar to safety fuse and may be used interchangeably with it. The fuse is a dark green cord 0.2 inches in diameter with a plastic cover, either smooth or with single painted bands around the outside at 1-foot or 18-inch intervals and double painted bands at 5-foot or 90-inch intervals, depending on the time of manufacture. These bands are provided for easy measuring purposes. The burning rate is approximately 40 seconds per foot, which permits the soldier firing the charge to reach a place of safety. The burning rate, however, must always be tested in the same manner as that of safety fuse, above. At arctic temperatures, the outside covering becomes brittle and cracks easily.

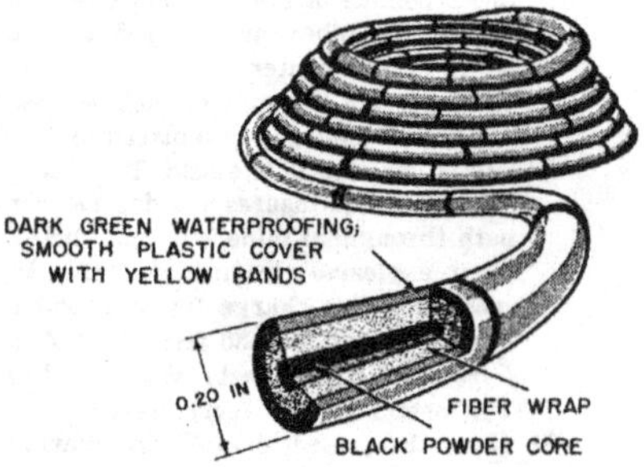

Figure 16. Time fuse M700.

c. Packaging.

(1) *Safety fuse.*

(*a*) 50-foot coil, 2 coils per package, and 30 packages (3000 feet) in a wooden box 24¾ x 15¾ x 12½ inches. The total package weight is 71.8 pounds.

(*b*) 50-foot coil, 2 coils per package, 5 packages sealed in a metal can, and 8 cans (4000 feet) per wooden box 30 x 14⅝ x 14⅝ inches. The total package weight is 93.6 pounds.

(*c*) 50-foot coil, 2 coils per package, and 60 packages (6000 feet) per wooden box 29 x 22 x 17 inches. The total package weight is 162 pounds.

(2) *Time fuse M700.* This is packed in 50-foot coils, 2 coils per package, 5 packages per sealed container, and 8 containers (4000 feet) per wooden box 30⅛ x 15⅛ x 14⅞ inches. The total package weight is 94 pounds.

30. Detonating Cord

a. Characteristics. Detonating cord consists of a core of PETN in a textile tube coated with a layer of asphalt. On top of this is an outer textile cover finished with a wax gum composition or plastic coating (fig. 17). It will transmit a detonating wave from one point to another at a rate of at least 5900 meters per second or about 19400 feet. Partially submerged water-soaked detonating cord will detonate if initiated from a dry end. Although it does not lose its explosive properties by exposure to low temperatures, the covering, becoming stiff, cracks when bent. Thus great care is required in using detonating cord primers in arctic conditions. Data on the types available is shown in table II.

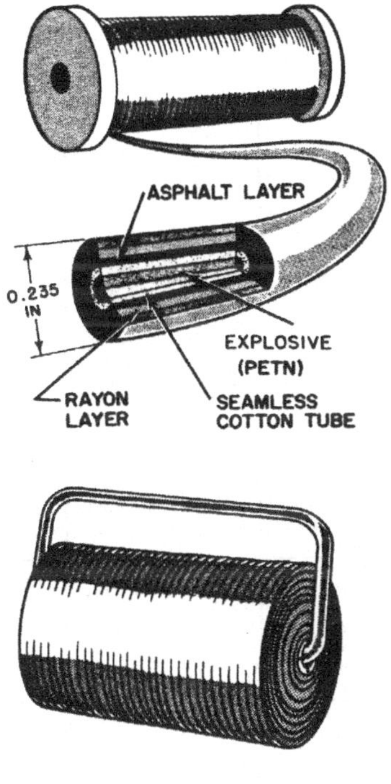

Figure 17. Reinforced pliofilm-wrapped detonating cord.

Table II. Detonating Cord Data

Nomenclature	Nom. dia. (in.)	Covering	Loading	Packing				
				Description	Dimension (inches)			Weight (lbs.)
					Length	Width	Height	
CORD, DETONATING: fuse, primacord (PETN) Type 1*.	0.210	Cotton with wax gum composition finish.	50 grn PETN/ft.	1. 1,000 ft/spool, 1 spool (1,000 ft)/wdn bx	10⅛	10	10	16
				2. 100 ft/spool, 25 spool (2,500 ft)/wdn bx	19¾	18⅛	8⅛	46
				3. 500 ft/spool, 1 spool/sealed can, 8 can (4,000 ft)/wdn bx	30	15	14⅝	105

Table II. Detonating Cord Data—Continued

Nomenclature	Nom. dia. (in.)	Covering	Loading	Packing				
				Description	Dimension (inches)			Weight (lbs.)
					Length	Width	Height	
				4. 500 ft/spool, 8 spool (4,000 ft)/wdn bx				
				5. 50 ft/spool, 100 spool (5,000 ft)/wdn bx	24	17	12	94
CORD, DETONATING: fuse, primacord (PETN) Type II*	0.216	Double cotton with wax gum composition finish.	50 grn PETN/ft.	100 ft/spool, 50 spool (5,000 ft)/wdn bx.	21	14¾	18⅝	111
CORD, DETONATING: fuse, primacord (PETN) (50-ft spool (spliced)).	0.210	Cotton with wax gum composition finish.	50 grn PETN/ft.	Packed as required				
CORD, DETONATING: fuse, primacord (PETN) inert.	0.210	Cotton with wax gum composition	Inert	Packed as required.				
CORD, DETONATING: reinforced, pliofilm-wrapped, waterproof, Type IV**.	0.235	Textile with plastic coating.	60 grn PETN/ft.	1. 1,000 ft/spool, 1 spool/crdbd bx, 3 bx/wtrprf lead foil env 1 env (3,000 ft)/wdn bx	33⅜	11¾	11¼	77
CORD, DETONATING: reinforced, pliofilm-wrapped, waterproof, Type IV**.				2. 500 ft/spool, 1 spool sealed can, 8 can (4,000 ft)/wdn bx				
CORD, DETONATING: reinforced, dummy.	0.235	Textile with plastic coating.	Inert	Packed as required				
CORD, DETONATING: waterproof, plastic outer covering (8-ft length) Type I**.	0.235	Textile with plastic coating.	60 grn PETN/ft.	200 length (1,600 ft)/wdn bx	26½	18	11	57

* Type designation in accordance with PA-PD-417, 7 May 1954.
** Type designation in accordance with MIL-C-17124A, 11 June 1959.

Abbreviations:

bx . . . . . box(es) dia . . . . . diameter ft . . . . . foot (feet) lb . . . . . pound(s) wdn . . . . wooden
crdbd . . . cardboard env . . . . . envelope grn . . . . . grain(s) nom . . . . nominal wtrprf . . . waterproof

b. Precautions in Use. The ends of detonating cord should be sealed with a waterproof sealing compound to keep out moisture when used to detonate underwater charges, or charges left in place several hours before firing. A 6-inch free end will also protect the remainder of a line from moisture for 24 hours. In priming, kinks or short bends, which may change the direction of detonation and thus cause misfires, should be avoided.

31. Blasting Caps

Blasting caps are used for initiating high explosives. They are designed for insertion in cap wells, and are also the detonating element in certain land mine firing devices. Special military blasting caps are designed to detonate the less sensitive explosives like TNT, military dynamite, and tetrytol. Commercial caps may be used to detonate more sensitive explosives like tetryl and commercial dynamite, in an emergency. Two commercial caps are required to detonate military explosives; however, there is also a priming problem as two caps will *not* fit into the standard threaded recess. Both military and commercial blasting caps, being extremely sensitive, may explode unless handled carefully. They must be protected from shock and extreme heat and not tampered with. Blasting caps must never be stored with other explosives, nor should they be carried in the same truck except in an emergency (para 141*b*(3)). Two types, electric and nonelectric, are used in military operations.

a. Electric Blasting Caps. These are used when a source of electricity, such as a blasting machine or a battery, is available. Two types are in use, military and commercial (fig. 18). Military caps are instantaneous, and the commercial, instantaneous and delay. Instantaneous caps include the M6, the special or military, and the commercial No. 6 and No. 8. No. 8 commercial delay caps (fig. 18) are issued from the first to the fourth delay ranging from 1:00 second to 1:53 seconds. When two or more of the special instantaneous caps are used, they should be of the same manufacture except for the M6 caps, which regardless of manufacturer, may be used interchangeably as they are all made to a single specification. All issue electric caps have lead wires of various lengths for connection in a circuit. The most commonly used are 12 feet long. Most all have a short circuiting shunt or tab, to prevent accidental firing, which must be removed before connection in a firing circuit. If the cap is issued without a shunt, which is sometimes the case with the M6, the bare ends of the lead wires must be twisted together to provide the shunting action. Data on electric blasting caps is shown in table III.

b. Nonelectric Blasting Caps. Nonelectric blasting caps (fig. 19) may be initiated by time blasting fuse, firing devices, and detonating cord. Because they are extremely difficult to waterproof, they should not be used with time blasting fuse to prime charges placed under water or in wet boreholes. If such be necessary, however, they should be moistureproofed with waterproof sealing compound. Those in use include the commercial No. 6 and No. 8 and the special or military types I (J-1 (PETN or RDX) and M7 (fig. 19). Special caps will detonate military explosives, and the commercial caps the more sensitive types. The latter, however, will detonate military explosives if used in pairs; but this presents a priming problem, as two caps will *not* fit into the standard threaded cap well. The M7 special caps are flared at the open end for easy insertion of the time fuse. Data on nonelectric caps is shown in table IV.

32. Priming Adapter M1A4

This is a plastic hexagonal-shaped device threaded to fit threaded cap wells and the M10 universal destructor. A shoulder inside the threaded end is large enough to accept blasting fuse and detonating cord but too small to permit passage of a blasting cap. The adapter is slotted longitudinally to permit easy and quick insertion of the electric blasting cap lead wires (fig. 20). The M1A4 replaces the M1A2 and M1A3 models, which have cylindrical bodies. The hexagonal M1A4 is more readily handled by men wearing arctic mittens.

33. Adhesive Paste, M1

This is a sticky, putty-like substances for attaching charges to vertical or overhead flat surfaces. It is useful in holding charges while

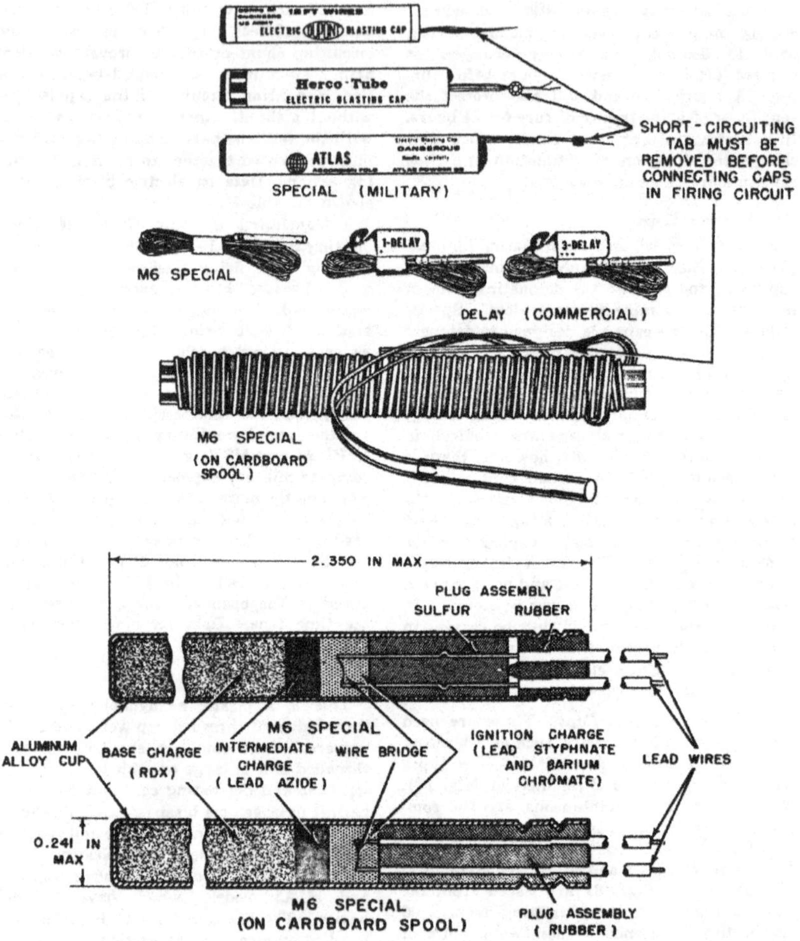

Figure 18. Electric blasting caps.

Table III. Electric Blasting Cap Characteristics

Nomenclature	Lead wire characteristics	Packing: Description	Dimensions (inches): Length	Dimensions (inches): Width	Dimensions (inches): Height	Weight (lbs.)
CAP, BLASTING: commercial, electric, No 6, instantaneous	Short lead, 4 ft through 10 ft	500/wdn bx	18	13⅝	12	48.6
CAP, BLASTING: commercial, electric, No 6, instantaneous	Medium lead, 12 ft through 40 ft	As required				
CAP, BLASTING: commercial, electric, No 6, instantaneous	Long lead, 50 ft through 100 ft	As required				
CAP, BLASTING: commercial	Short lead, 4 ft through 10 ft	1. 70/ctn. 5 ctn (350 cap)/wdn bx	15	12½	6½	22.2
CAP, BLASTING: commercial electric, No 8, instantaneous	Medium lead, 12 ft through 40 ft	As required				
CAP, BLASTING: commercial electric, No 8, instantaneous	Long lead, 50 ft through 100 ft	As required				
CAP, BLASTING: electric No 8, strength	Lead 6 ft long, copper tinned	50/ctn				
CAP, BLASTING: electric, No 8, strength	Lead 6 ft long	70/ctn	13¾	7¼	4¾	8
CAP, BLASTING: electric, No 8, strength	Lead 30 ft long, copper tinned	25/ctn				
CAP, BLASTING: electric, No 8, 1st delay (approx 1.00 sec)	Lead 12 ft long	500/wdn bx	19½	13½	9¼	30
CAP, BLASTING: electric, No 8, 2nd delay (approx. 1.18 sec)	Lead 12 ft long	500/wdn bx	19½	13½	9¼	30
CAP, BLASTING: electric, No 8, 3rd delay (approx. 1.35 sec)	Lead 12 ft long	500/wdn bx	19½	13½	9¼	30.5
CAP, BLASTING: electric ,No. 8, 4th delay (approx. 1.53 sec)	Lead 12 ft long	500/wdn bx	19½	13½	9¼	30.0
CAP, BLASTING: electric, high strength	Lead 6 ft long, copper tinned	1. 50/ctn 2. 70/ctn	9 14½	4⅛ 7⅛	3 3¾	2.3 8.0
CAP, BLASTING: electric, high strength	Lead 9 ft long, copper tinned	50/ctn	7¼	4⅝	3¼	4.0
CAP, BLASTING: electric, low strength	Lead 6 ft long, copper tinned	50/ctn	6½	6	2⅞	2.0
CAP, BLASTING: electric, inert	Various long lead wires	As required				
CAP, BLASTING: special, electric	Lead 12 ft long	1. 1/chipbd pkg. 50 pkg/fbrbd bx, 10 bx (500 cap)/wdn bx 2. As required	28¼ 17½	15¼ 12½	11¾ 11	76.5 51.0

Table III. *Electric Blasting Cap Characteristics*—Continued

Nomenclature	Lead wire characteristics	Packing				
		Description	Dimensions (inches)			Weight (lbs.)
			Length	Width	Height	
CAP, BLASTING: special, electric, M6	Lead 12 ft long, copper tinned	1/crdbd spool, 6 spool/ctn, 1 ctn/water-prf bg, 25 bg/fbrbd bx, 6 fbrbd bx (900 cap)				

Abbreviations:
bg . . . . . bag(s) chipbd . . . chipboard ctn . . . . . carton(s) ft . . . . . . foot(feet)
bx . . . . . box(s) crdbd . . . . cardboard fbrbd . . . . fiberboard lb . . . . . .pound(s)

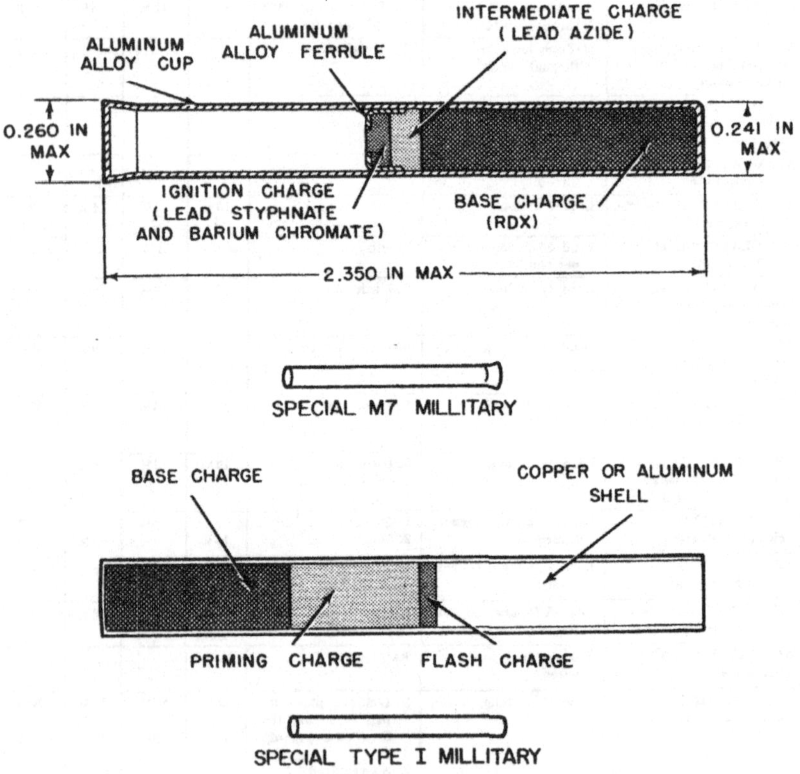

Figure 19. Military nonelectric blasting caps.

Table IV. Nonelectric Blasting Cap Characteristics

Nomenclature	Packing				
	Description	Dimensions (inches)			Weight (lbs.)
		Length	Width	Height	
CAP, BLASTING: nonelectric, No 6, instantaneous.	1. 5,000/wdn bx	20⅛	9⅞	8¾	29
	2. 100/ctn, 10ctn/fbrbd bx, 1 bx/wtrprf bag, 5 bag (5,000 cap)/wdn bx				
CAP, BLASTING: nonelectric, No 8, instantaneous.	1. 100/ctn, 50 ctn (5,000 cap)/wdn bx	18½	13	7¾	50
	2. As required				
CAP, BLASTING: special, nonelectric (Type I(J-1)) (PETN or RDX).	1. 50/mtl can, 20 can/fbrbd ctn, 5 ctn (5,000 cap)/wdn bx	24¾	17¼	14¼	82
	2. 100/ctn, 10 ctn/crdbd bx, 10 bx (10,000 cap)/inner wtrprf pkg, 1 inner pkg in sawdust/outer wtrprf pkg/wdn bx	22¾	17⅛	14	66
	3. As required				
CAP, BLASTING: special, nonelectric, M7.	6/pprbd ctn, 1 ctn/wtrprf bag, 50 bag/fbrbd bx, 12 bx (3,600 cap)/wdn bx				
CAP, BLASTING: tetryl, nonelectric, type A.	100/ctn, 50 ctn (5,000 cap)/wdn bx	19	16	8	65
CAP, BLASTING: nonelectric, inert.	As required				

Abbreviations:

bx . . . . . box(es) ctn . . . . . carton lb . . . . . . pound pkg . . . . package(s) wdn . . . . wooden

crdbd . . . . cardboard fbrbd . . . . fiberboard mtl . . . . . metal pprbd . . . . paperboard wtrprf . . waterproof

tying them in place or, under some conditions, for holding without tying. It will not adhere satisfactorily to dirty, dusty, wet or oily surfaces; becomes stiff and hard and loses its adhesiveness at subzero temperatures; is softened by water; and becomes useless if wet.

34. Waterproof Sealing Compound

This is used to waterproof the connection between the time blasting fuse and a nonelectric blasting cap and to moistureproof primed dynamite. It does not make a permanent waterproof seal and must not be submerged in water unless the charge is to be fired immediately.

35. Cap Crimper

The M2 cap crimper (fig. 21) is used to squeeze the shell of a nonelectric blasting cap around time fuse, a standard base, or detonating cord securely enough to keep it from being pulled off but not tightly enough to interfere with the burning of the powder train in the fuse or the detonation of the detonating cord. The M2 crimper forms a water resistant groove completely around the blasting cap; however, sealing compound should be applied to the crimped end of the blasting cap for use under water. The rear portion of the jaws is shaped and sharpened for cutting fuse and detonating cord. One leg of the handle is pointed for use in punching fuse wells in explosive materials for the easy insertion of blasting caps. The other leg has a screwdriver end. Cap crimpers being made of a soft nonsparking metal (but they will conduct electricity), must not be used as pliers for any purpose, as this damages the crimping surface. Also the cutting jaws must be kept clean and be used only for cutting fuse and detonating cord.

36. Galvanometer

The galvanometer is an instrument used in testing the electric firing system to check the continuity of the circuit (the blasting cap, firing wire, wire connections, and splices) in

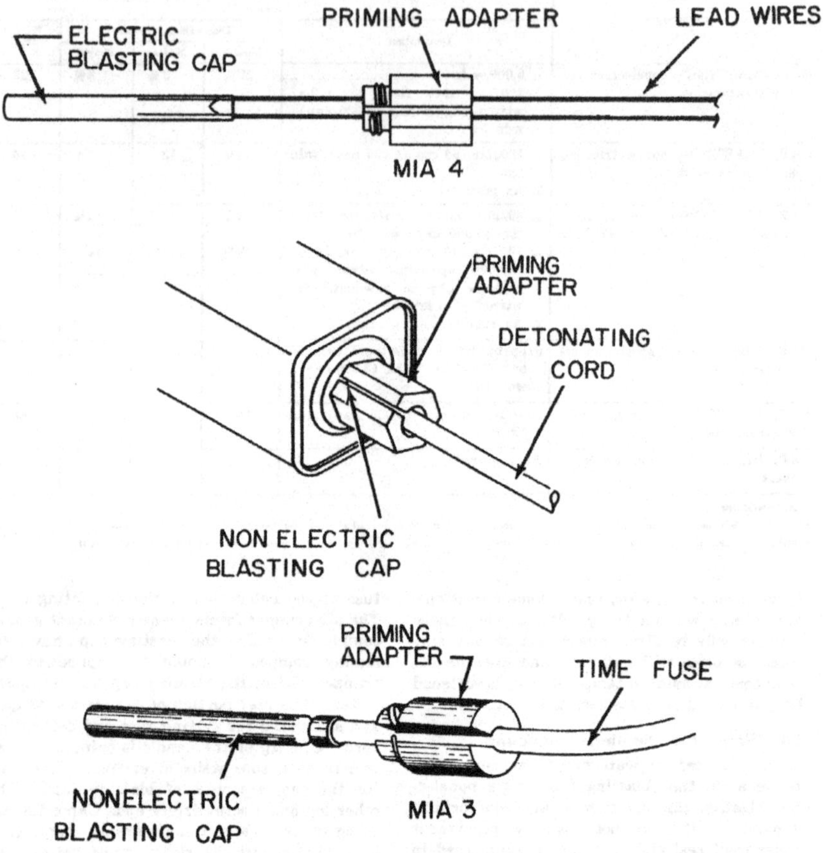

Figure 20. Priming adapters.

order to reduce the possibility of misfires (fig. 22). Its components include an electromagnet, a small special silver-chloride dry cell battery, a scale, and an indicator needle. When the two external terminals are connected in a closed circuit, the flow of current from the dry cell moves the needle across the scale. The extent of the needle deflection depends on the amount of resistance in the closed circuit and on the strength of the battery. The galvanometer must be handled carefully and kept dry. It should be tested before using by holding a piece of metal across its two terminals. If this does not cause a wide deflection of the needle (23 to

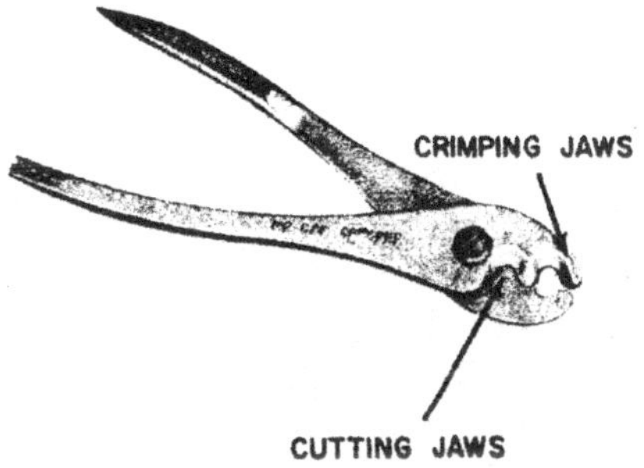

Figure 21. M2 Cap crimper.

25 units) the battery is weak and should be replaced. Being delicate, the instrument must not be opened except to replace a weak cell. When used in a cold climate, the galvanometer should be protected from freezing by keeping it under the clothing near the body, as dry cell batteries tend to cease functioning at temperatures below 0°F.

CAUTION: Only the special silver-chloride dry cell battery BA 245/U, which produces only 0.9 volts, is to be used in the galvanometer, as other batteries may produce sufficient voltage to detonate electric blasting caps. Because of the tendency to corrode, the battery should be removed from the galvinometer when it is not to be used for extended periods.

37. Blasting Machine

a. Ten-Cap Blasting Machine. This is a small electric impulse-type generator that produces adequate current (45 volts) to initiate 10 electric caps connected in series if the handle is rotated to the end of its travel. It weighs approximately 5 pounds (fig. 23). The operation is as follows:

(1) Try the machine to see whether or or not it works properly. Operate it several times until it works smoothly before attaching the firing wires.

(2) Fasten the firing wires tightly to the terminals.

(3) Insert the handle.

(4) Insert the left hand through the strap and grasp the bottom of the machine.

(5) Grasp the handle with the right hand and *turn it vigorously clockwise as far as possible.*

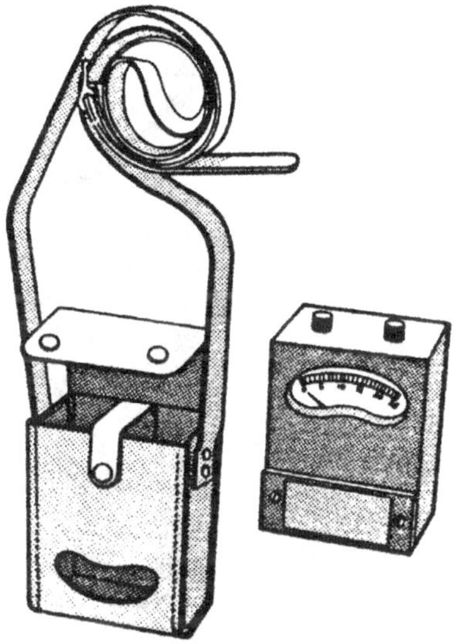

Figure 22. Galvanometer.

b. Thirty-Cap Blasting Machine. This device fires 30 electric caps connected in series. It weighs about 20 pounds. To operate:

(1) Raise the handle to the top of the stroke.

(2) Push the handle down quickly as far as it will go.

c. Fifty-Cap and One-Hundred Cap Blasting Machines.

(1) The 50-cap machine fires 50 electric caps connected in series (fig. 23). It weighs about 20 pounds. Operation is as follows:

(*a*) Raise the handle to the top of its stroak.

(*b*) Push the handle down quickly as far as it will go.

(2) The 100-cap machine is similar to the

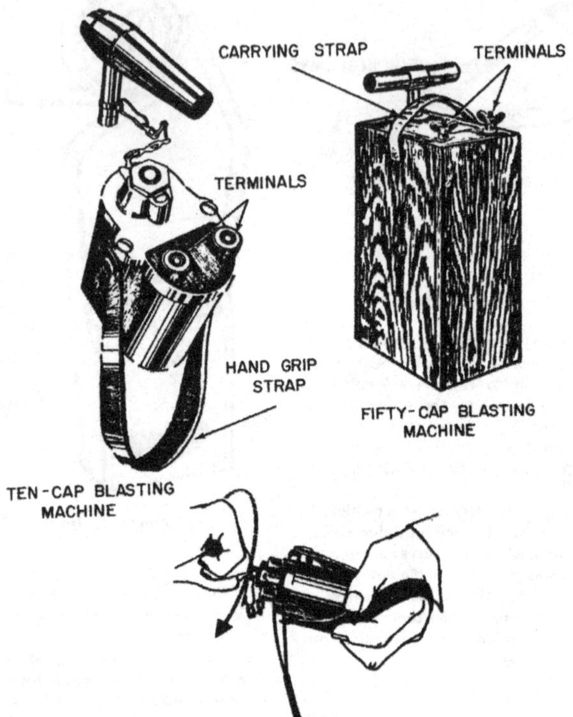

Figure 23. Blasting machines.

50-cap machine except for size and weight and is operated in the same manner. Both are adequate for firing their rated capacity of electric blasting caps connected in series

38. Firing Wire and Reel

a. Types of Firing Wire. Wire for firing electric charges is issued in 500-foot coils. It is the two-conductor, No. 18 AWG plastic-covered or rubber-covered type. It is carried on the reel unit RL39A, described below. Single-conductor No. 20 AWG annunciator wire in 200-foot coils is issued for making connections between blasting caps and firing wire. WD-1/TT communication wire may also be used. However, it has a resistance of about

40 ohms per 1000 feet, which increases the power requirement (table XXI, app E).

b. Reels.

(1) *RL39A*. This consists of a spool that accommodates 500 feet of wire, a handle assembly, a crank, an axle, and two carrying straps (fig. 24). The fixed end of the wire is extended from the spool through a hole in the side of the drum and fastened to two brass thumbnut terminals. The carrying handles are made of two U-shaped steel rods. A loop at each end encircles a bearing assembly, which is a brass housing with a steel center to accommodate the axle. The crank is riveted to one end of the axle and a cotter pin is placed in the hole at the other to hold the axle in place.

(2) *500-foot reel with detachable handles.* This is a metal drum mounted on an axle to which two detachable D-shaped handles are fastened. The arm with the knob on the side of the drum is used for cranking (fig. 25).

Figure 24. Firing wire reel.

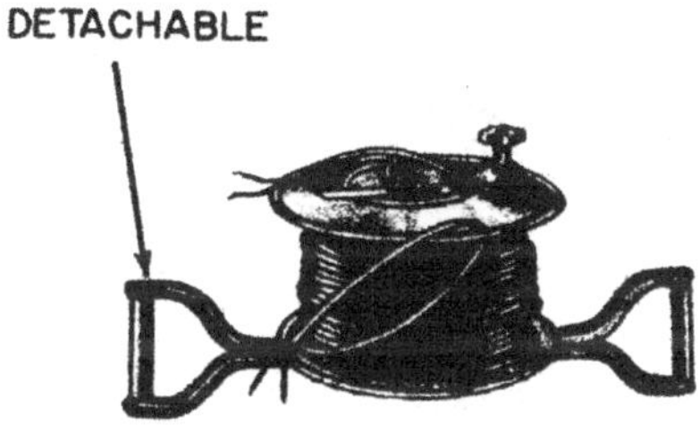

Figure 25. Reel with detachable handles.

(3) *1000-foot reel.* This is similar to (2) above, except that it has a capacity of 1000 feet of firing wire.

39. Detonating Cord Clip

The M1 detonating cord clip (fig. 26) is used to hold together two strands of detonating cord either parallel or at right angles to each other. Connections are made more quickly with these clips than with knots. Also, knots may loosen and fail to function properly if left in place any length of time. Joints made with clips are not affected by long exposure.

a. Branch Line Connections. Branch lines of detonating cord are connected by clipping the branch line with the U-shaped trough of the clip, and the main line with the tongue of the clip, as shown in figure 26.

b. Connecting Two Ends. Ends of detonating cord are spliced by overlapping them about

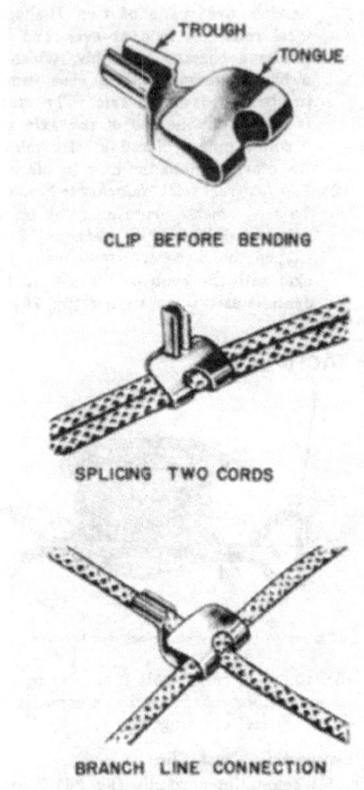

Figure 26. M1 Detonating cord clip.

12 inches, using two clips, one at each end of the overlap, and bending the tongues of the clips firmly over both strands. The connection is made secure by bending the trough end of the clip back over the tongue (fig. 26).

40. Firing Devices and Other Accessory Equipment

a. M10 Universal High Explosive Destructor. The M10 destructor is a high explosive charge in an assembled metal device initiated by means of blasting caps or mine activators with standard firing devices. The destructor has booster cups containing tetryl pellets. The chief function of the destructor is the conversion of loaded projectiles and bombs to improvised demolition charges and the destruction of abandoned ammunition (fig. 27).

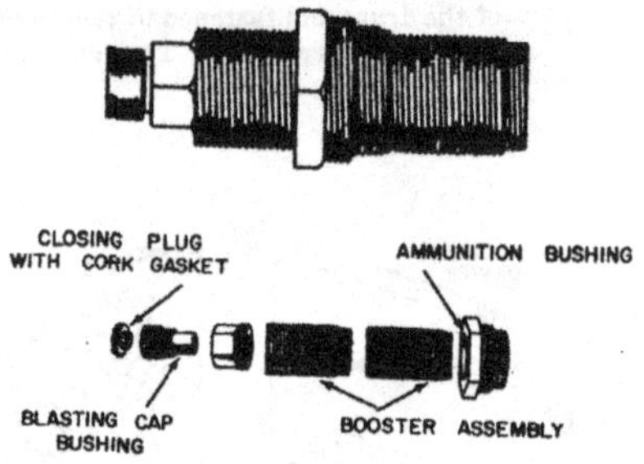

Figure 27. M10 destructor.

b. M19 Explosive Destructor. This device (fig. 28) consists of an explosive-filled cylindrical body with a removable pointed ogive, which may be discarded if not needed. This destructor may be primed with a delay detonator, delay firing device with a special blasting cap, a nonelectric special blasting cap initiated by time blasting fuse or detonating cord, or an electric special blasting cap. The cap well on each end is threaded to receive the standard base coupling or a priming adapter. This device is particularly suitable for use with the dust initiator, described in paragraph 8, appendix E, and similar charges.

c. M1 Concussion Detonator. The M1 concussion detonator is a mechancal firing device actuated by the concussion wave of a nearby blast (fig. 29). It fires several charges simultaneously without connecting them with wire or detonating cord. A single charge fired in any way in water or air will detonate all charges primed with concussion detonators within range of the main charge or of each other (table V). Detonators frequently function at ranges greater than those in table V,

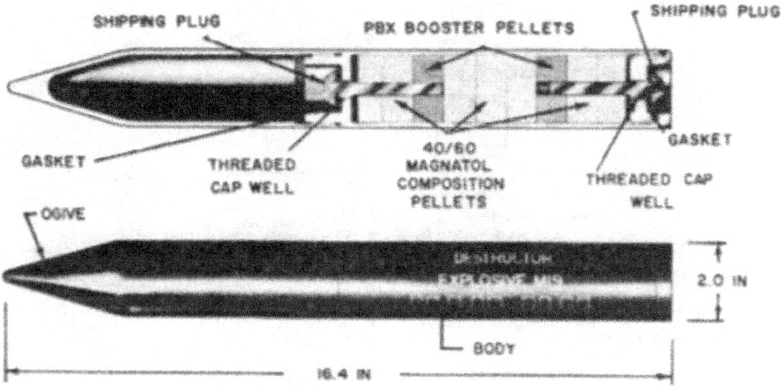

Figure 28. M19 explosive destructor.

but their reliability is then not assured. They should not be used in surf at depths greater than 15 feet, as they function by hydrostatic pressure at a depth of 25 feet. Further, if the salt delay pellet is crumbled due to long storage, the detonator should *not* be used on under-water charges.

Table V. Operating Range of Concussion Detonators

Initiating charge (lb)	In water: Depth of water (ft)	In water: Recommended range (ft)		In air: Recommended range (ft)
			P = 99%	P = 99%
0.5	2	10	—	—
0.5	4	50	—	—
0.5	6	80	—	—
0.5	8	80	—	—
2.5	—	—	12.5	10.8
2.5	2	20	—	—
2.5	4	80	—	—
2.5	6	90	—	—
2.5	8	150	—	—
5	—	—	14.1	11.5
10	—	—	18.8	15.7
15	—	—	21.5	18.0
20	—	—	25.2	21.2
20	2	20	—	—
20	4	80	—	—
20	6	180	—	—
20	8	260	—	—

P=Probability that detonator at indicated distance will be functioned by initiating charge.

d. M1A1 15-Second Delay Friction Detonator.

(1) *Characteristics* (fig. 30). This device consists of a pull friction fuse igniter, 15-second length of fuse, and blasting cap. The blasting cap is protected by a cap screwed on the base.

(2) *Installing.*

(*a*) Unscrew cap protector from base.

(*b*) Secure device in charge.

e. M1A2 (M1E1) 15-Second Delay Percussion Detonator.

(1) *Characteristics* (fig. 31). This item consists of a firing pin assembly joined to a delay housing and primer holding assembly.

(2) *Installing.*

(*a*) Remove cap protector.

(*b*) Screw device into threaded cap well.

f. M2A1 (M2E1) 8-Second Delay Percussion Detonator.

(1) *Characteristics* (fig. 32). Except for the delay period, marking, and shape of the pull ring, the 8-second delay percussion detonator is identical with *e*, above, in construction, functioning, and installing.

(2) *Installing.*

(*a*) Remove cap protector.

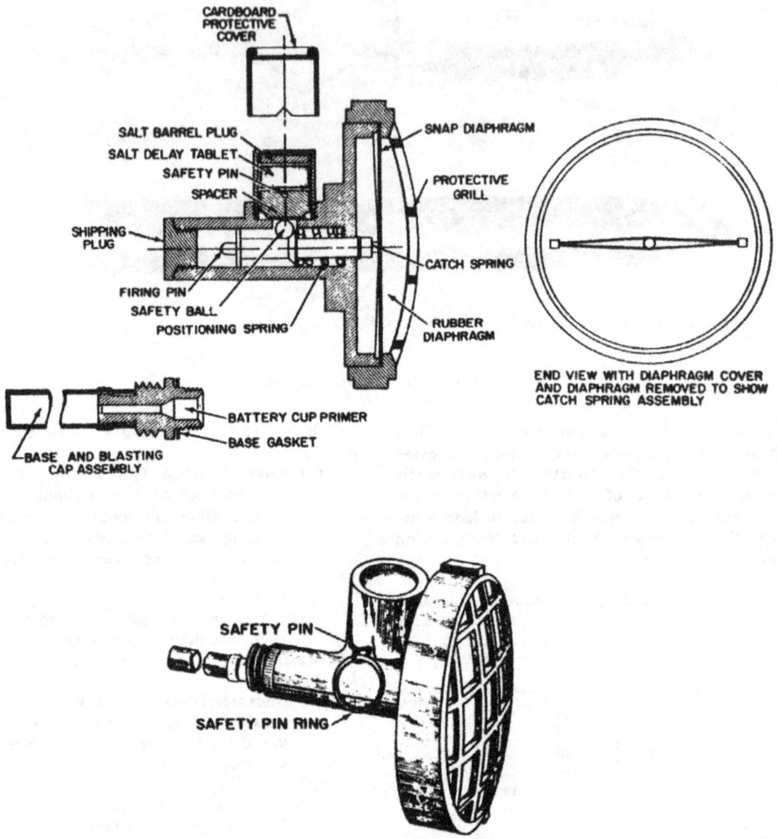

Figure 29. M1 Concussion detonator.

(*b*) Screw device into threaded cap well.

g. M2 8-Second Delay Friction Detonator.

(1) *Characteristics* (fig. 33). This device consists of a friction-type fuse lighter, an 8-second length of fuse, and a blasting cap. The blasting cap is protected by a cap screwed on the base.

(2) *Installing.*

(*a*) Unscrew cap protector from base.

(*b*) Secure device in charge.

h. M1 Delay Firing Device.

(1) *Characteristics* (fig. 34).

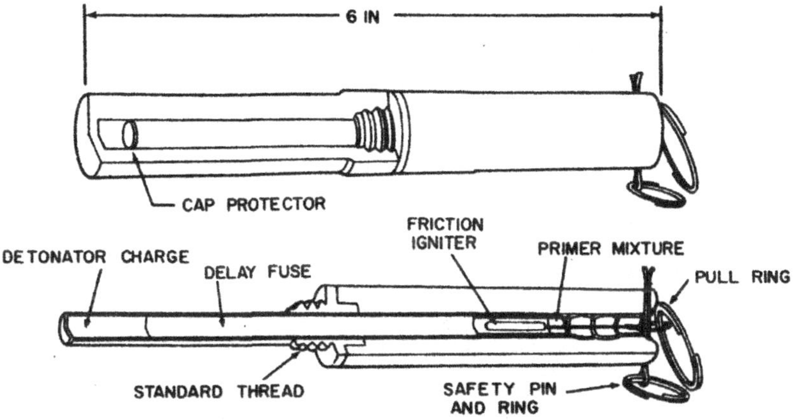

Figure 30. M1A1 15-second delay friction detonator.

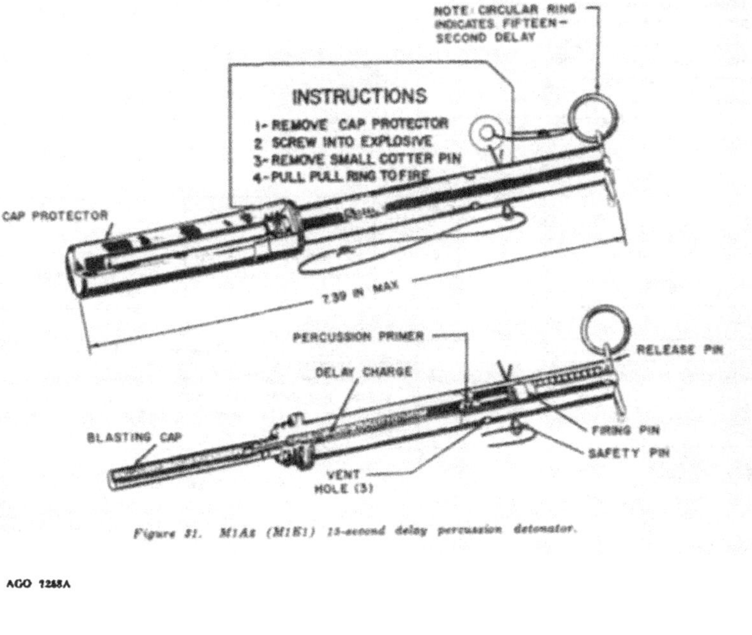

Figure 31. M1A2 (M1E1) 15-second delay percussion detonator.

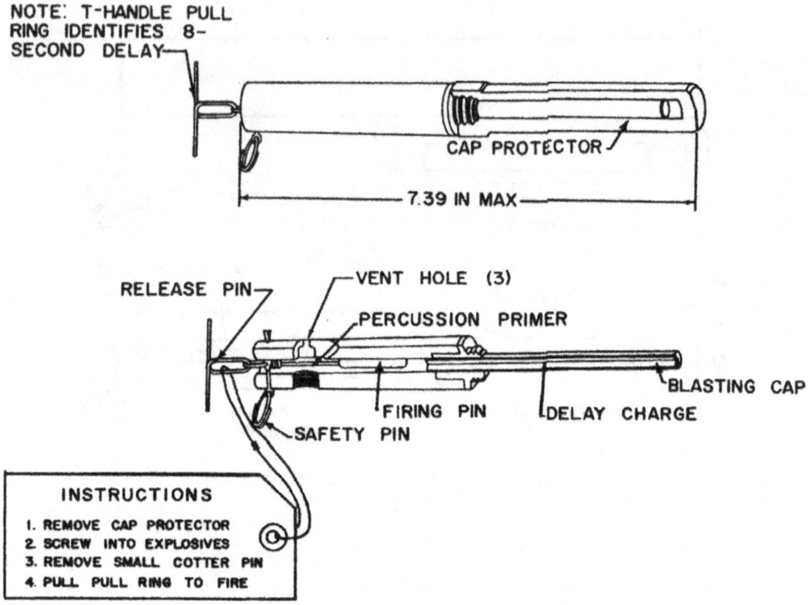

Figure 32. M2A1 (M2E1) 8-second delay percussion detonator.

Case	Color	Dimensions		Internal action	Delay
		D	L		
Copper and brass	Natural metal	7/16 in	6¼ in	Mechanical with corrosive chemical release.	1 min to 23 days, identified by color of safety strip.

Safety	Packaging
Colored metal strip inserted in slot above percussion cap.	10 units—2 red, 3 white, 3 green, 1 yellow, and 1 blue—and a time delay table packed in paperboard carton, 10 cartons in fiberboard box, and 5 boxes in wooden box.

(2) *Installing.*

(*a*) *Select device of proper delay* (table VI).

(*b*) Insert nail in inspection hole to make sure that firing pin has not been released. If the firing pin has been released, the nail cannot be pushed through the device.

Table VI. Temperature Corrections for M1 Delay Firing Device

Temp (deg F)	Black		Red		White		Green		Yellow		Blue		Temp (deg C)
	OM	ST	OM	ST	OM	ST	OM	ST	OM	ST	OM	ST	
-25			8.5 hr	3.3 hr	3 da	1.3 da							-32
0	8 hr	2.5 hr	45 min	20 min	17.5 hr	8 hr	2.6 da	1.2 da	8.5 da	3.8 da	23 da	10 da	-18
+25	36 min	16 min	25 min	11 min	5.5 hr	2.5 hr	17 hr	8 hr	2.0 hr	20 hr	5.0 da	2.2 da	-4
50	15 min	7 min	17 min	8 min	2 hr	55 min	6 hr	2.7 hr	14 hr	6.0 hr	1.3 da	14 hr	+10
75	9 min	4 min	15 min	7 min	1 hr	27 min	2.5 hr	70 min	5.5 hr	2.5 hr	11.5 hr	5 hr	24
100	5 min	2.0 min	8 min	3.5 min	32 min	14 min	70 min	30 min	2.5 hr	65 min	5.2 hr	2.3 hr	38
125	4 min	1.5 min	5 min	2 min	20 min	9 min	35 min	15 min	80 min	36 min	2.5 hr	1.1 hr	52
150	3 min	1 min	4 min	1.5 min	15 min	6 min	20 min	9 min	46 min	21 min	80 min	36 min	66

OM—Most likely delay if two devices are used in the same charge. If only a single device is used, this value should be increased approximately 15 percent.

ST—Reasonable safe time. Delays of less than this value should not occur more than one in a thousand.

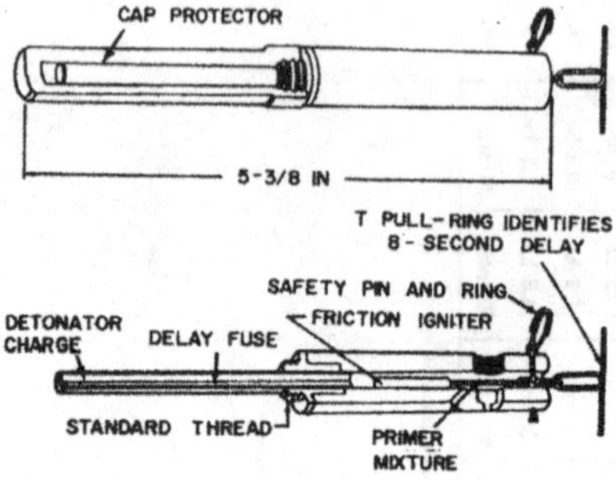

Figure 33. M2 8-second delay friction detonator.

(c) Remove protective cap from base.

(d) With crimper attach nonelectric blasting cap to base. *Crimper jaws should be placed no further than ¼ inch from open end of blasting cap.*

(e) Secure firing device in charge.

(f) Crush glass ampoule between thumb and fingers.

(g) Remove safety strip.

Caution: **If safety strip does not remove easily, remove and discard device.**

i. *M1A1 Pressure Firing Device.*

(1) *Characteristics* (fig. 35).

Case	Color	Dimensions		Internal action	Operating pressure
		D	L		
Metal	OD	⅝ in	2 ¾ in	Spring-driven striker with trigger pin and keyhole slot release.	20 lb or more.

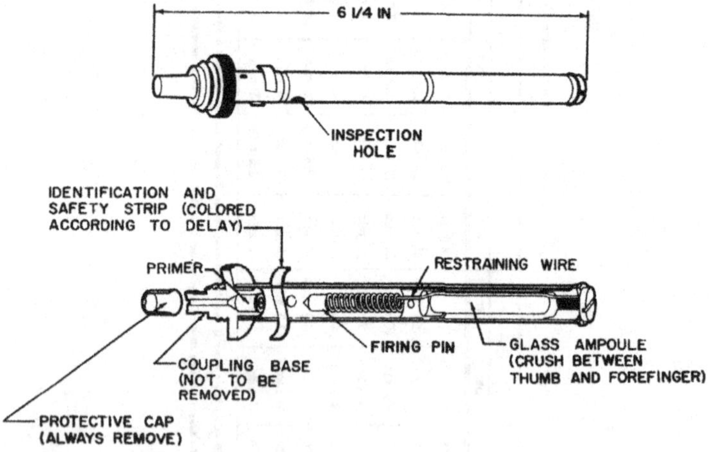

Figure 34. M1 delay firing device.

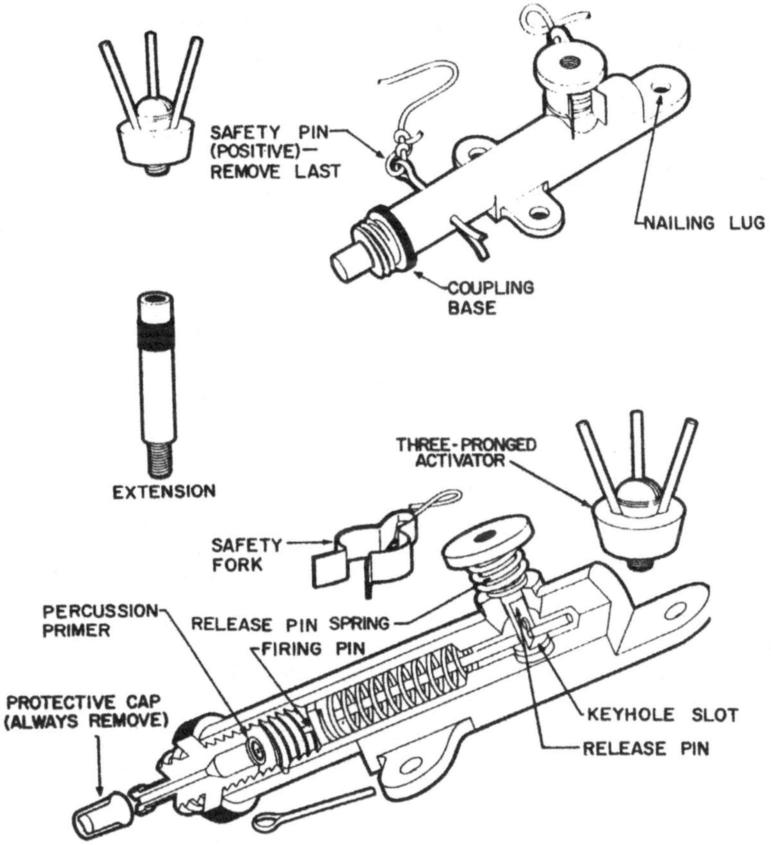

Figure 35. M1A1 pressure firing device.

Safeties	Accessories	Packaging
Safety fork and positive safety pin.	3-pronged pressure head and extension rod.	Five units, with percussion caps packed in cardboard carton. Fifty cartons shipped in wooden box.

(2) *Installing.*

(*a*) Remove protective cap from base and crimp on a nonelectric blasting cap. *Crimper jaws should be placed no farther than ¼ inch from open end of blasting cap.*

(*b*) Assemble 3-pronged pressure head and extension rod and screw in top of pressure cap, if needed.

(*c*) Attach firing device assembly to charge.

(*d*) Remove safety fork first, and positive safety last.

j. M1. Pull Firing Device.

(1) *Characteristics* (fig. 36).

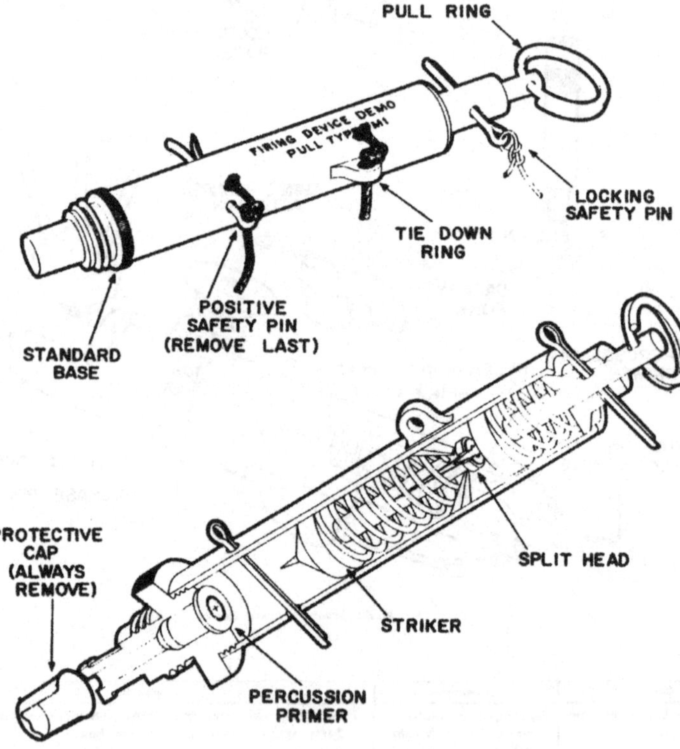

Figure 36. M1 pull firing device.

Case	Color	Dimensions		Internal action	Operating force
		D	L		
Metal	OD	9/16 in	3 5/16 in	Mechanical with split-head striker release.	3 to 5 lb pull on trip wire.

Safeties	Packaging
Locking and positive safety pins.	Five units complete with percussion caps and two 80-ft spools of trip wire are packed in chipboard container. Forty chipboard containers are packed in wooden box.

2. *Installing.*
   (*a*) Remove protective cap.
   (*b*) With crimpers, attach nonelectric blasting cap to standard base. *Crimper jaws should be placed no farther than ¼ inch from open end of blasting cap.*
   (*c*) Attach firing device assembly to charge.
   (*d*) Attach anchored pull wire.
   (*e*) Remove *locking safety pin first,* and *positive safety pin last.*

k. M3 Pull-Release Firing Device.

(1) *Characteristics* (fig. 37).

Case	Color	Dimensions		Internal action	Operating pressure
		D	L		
Metal	OD	9/16 in	4 in	Mechanical with spreading striker head release.	Direct pull of 6 to 10 lb.

Safeties	Packaging
Locking and positive safety pins.	Five units with two 80-ft spools of trip wire in carton, and 5 cartons packed in wooden box.

(2) *Installing.*
   (*a*) Remove protective cap.
   (*b*) With crimpers, attach blasting cap to standard base. *Crimper jaws should be placed no farther than ¼ inch from open end of blasting cap.*
   (*c*) Attach firing device to anchored charge (must be firm enough to withstand pull of at least 20 pounds).
   (*d*) Secure one end of trip wire to anchor and place other end in hole in winch.
   (*e*) With knurled knob, draw up trip wire until locking safety pin is pulled into wide portion of safety pin hole.
   (*f*) Remove *locking safety pin first* and *positive safety pin last.*

l. M5 Pressure-Release Firing Device.

(1) *Characteristics* (fig. 38).

Case	Color	Dimensions			Internal action	Initiating action
		L	W	Ht		
Metal	OD	1 ¾ in	15/16 in	11/16 in	Mechanical with hinged plate release.	Removal of restraining weight—5 lb or more.

Safeties	Packaging
Locking safety pin and hole for improvised positive safety pin.	Four firing devices complete and four plywood pressure boards in paper carton. Five cartons are packaged in fiber board box and ten of these in wooden box.

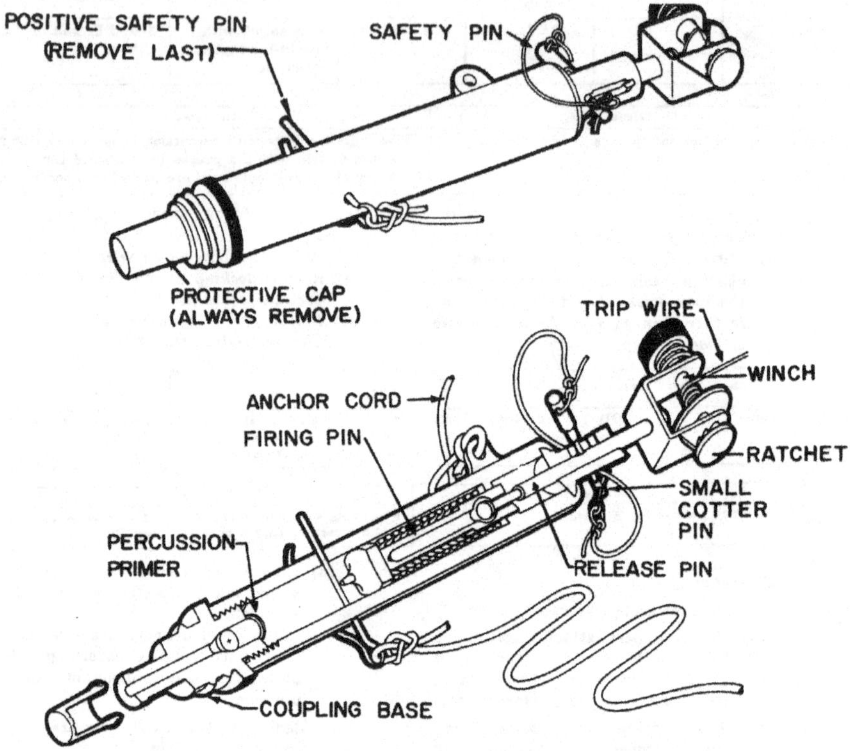

Figure 37. M3 pull-release firing device.

(2) *Installing.*

(*a*) Insert a length of 10-gage wire in interceptor hole.

(*b*) Bend slightly to prevent dropping out.

(*c*) Remove small cotter pin from safety pin.

(*d*) Holding release plate down, replace locking safety pin with length of 16 or 18 gage wire. Bend wire slightly to prevent dropping out.

(*e*) Remove protective cap from base and with crimpers, attach blasting cap. *Crimper jaws should be placed no farther than 1/4 inch from open end of blasting cap.*

(*f*) Secure firing device assembly in charge.

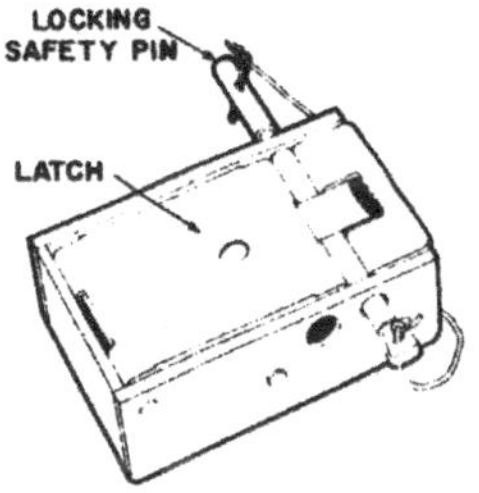

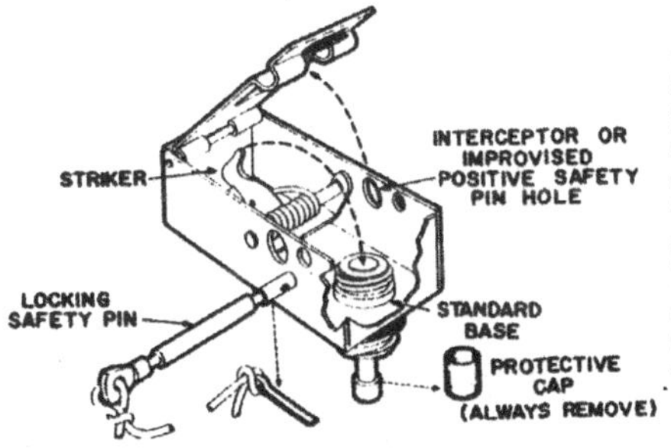

Figure 38. M5 pressure-release firing device.

(*g*) Emplace charge and set the restraining weight (5 pounds or more) on top of the firing device.

(*h*) Slowly and carefully, without disturbing the restraining weight, remove the improvised locking safety pin first and the improvised positive safety pin from the interceptor hole last. The pins should remove easily if the restraining weight is adequate and positioned properly.

m. M2 Weatherproof Lighter. This device was designed as a positive method of lighting time blasting fuse (fig. 39). It operates effectively under all weather conditions—even under water if it is properly waterproofed. A pull on the striker retaining pin causes the striker to hit the primer, igniting the fuse. A sealing compound is applied between the fuse and the lighter to retard any flash that may come from lighting the fuse.

n. M60 Weatherproof Fuse Lighter. This device is designed to ignite blasting fuse in all sorts of weather conditions and under water if waterproofed. The fuse is inserted into a fuse retainer and sealed and weatherproofed by means of two rubber washer seals (fig. 40). A pull on the pull ring releases the striker assembly, allowing the firing pin to drive against the primer, which ignites and initiates the fuse. For further information, see paragraph 43*i*.

o. Computing Tape. The demolition charge computing tape (fig. 41), provides a rapid method of calculating the weight of TNT (in pounds) needed to carry on a demolition project. It combines in an abbreviated form most of the formulas and tables provided in this text. The assembly consists of two 6-foot flexible steel spring retractable tapes in joined metal housings. The two tapes have a total of five sets of markings. A rigid embossed scale is mounted on one side of the housing. The scales are—

(1) *First tape (breaching and pressure scales).* The upper side of this tape indicates the pounds of TNT required to breach concrete, masonry, timber, or earthen walls, making allowances for the tamping and placement of charges. The weight is read directly to the right of the mark that indicates the thickness of the wall or obstacle. The lower side of the tape has information on breaching concrete beams, roadways, and bridge spans. It is used to measure the thickness of the target or element. The weight of the charge may be read directly from the tape without consideration of the actual dimensions of the target.

(2) *Second tape (steel- and timber-cutting scales).* This tape contains the requirements for cutting steel and timber construction materials. One side shows the weight of TNT needed for

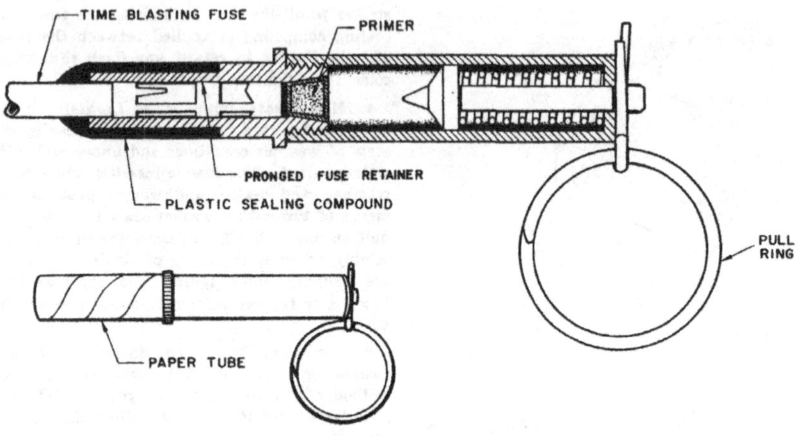

Figure 39. M2 weatherproof fuse lighter.

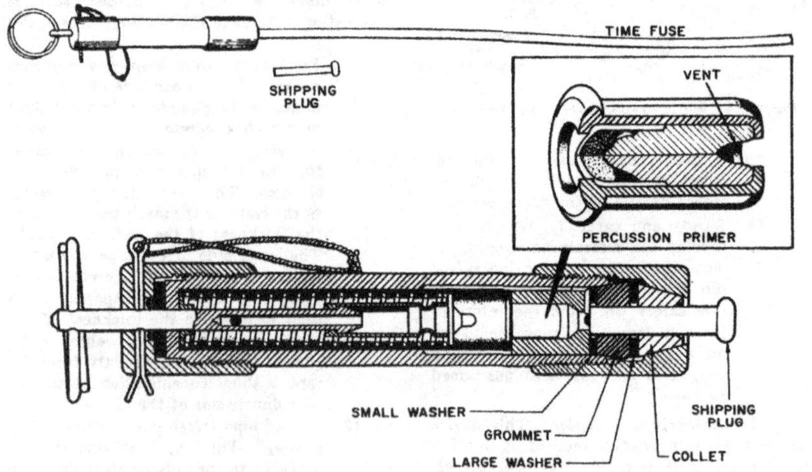

Figure 40. M60 weatherproof fuse lighter.

Figure 41. Computing tape.

cutting timber for both internal and external placement. The reverse side has a rule for the calculation of the cross-sectional area of steel members and also the formulas for cutting steel.

(3) *Bar and rod-cutting scale.* The small scale on the exterior of the case is used for making calculations for the cutting of rods, bars, chains, and cables. The number of pounds of TNT needed for cutting is read directly from the scale.

p. Demolition Card. This pocket sized card (GTA 5–10–9, May 65) gives data in tabulated form for the calculation of pressure, timber-cutting, steel-cutting, breaching, and cratering charges.

q. Rivet-Punching Powder-Actuated Driver (Ram-Set Gun).

(1) *Description.* This is a riveting machine powered by the gases generated by a fired cartridge (fig. 42). It is hand-operated, air-cooled, and feeds from a magazine with a 10-cartridge fastener unit capacity. It operates effectively under water. The waterproofed fastener unit has a sharp point and a coarsely knurled body to provide maximum holding power in light steel, softer metals, concrete, and heavy wood. The sabot, an annular threaded unit, screws on the rear of the fastener to guide it in ejection, acts as a atop-shoulder, and provides additional bearing on the penetrated material. The cartridge case is a specially wadded caliber .38 steel case. A manual device is provided for cocking the driver under water.

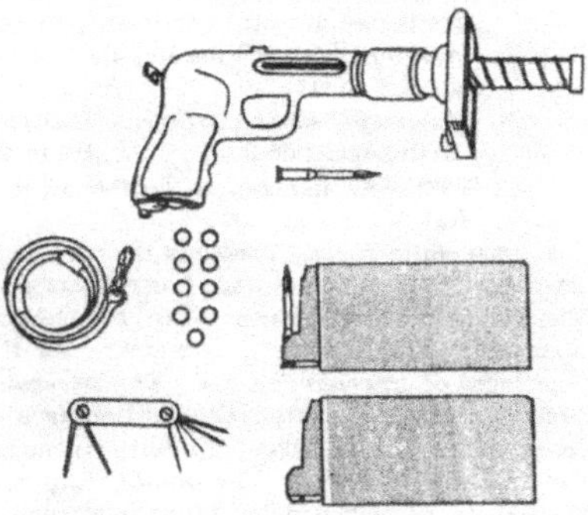

Figure 42. Ram-set gun.

(2) *Operation.* The firing of the cartridge propels the fastener and sabot into the target. The fastener acts as a rivet for attaching charges to steel, concrete, or wooden targets. The device is especially useful where working space is severely limited and for underwater work. ***Do not fire the gun into explosive or immediately adjacent to exposed explosives.***

r. Earth Augers and Diggers. Two types of earth augers, hand-operated and motorized, are used for boring holes for the placement of cratering charges and bridge-abutment demolition charges. Motorized earth augers will bore holes to a depth of approximately 9 feet. Boring speed depends on the type and consistency of the soil, being most rapid in light earth or loam. Earth augers and diggers cannot be used satisfactorily in soil containing large rocks.

(1) *Hand-operated posthole auger.* The 10-inch posthole auger (fig. 43) is capable of boring a hole large enough for the 40-pound ammonium nitrate cratering charge and other charges of equal size. The extension handle permits boring as deep as 8 feet.

(2) *Posthole digger.* This tool (fig. 43) has two concave metal blades on hinged wooden handles. The blades are forced into the earth and the soil is removed by lifting and pulling the handles apart.

(3) *Motorized earth auger.* Motorized earth augers drill hole 8, 12, 16, or 20 inches in diameter to depths up to 8 feet.

s. Pneumatic Tools. These are the rock drill, pavement breaker, and wood-boring machine. The *rock drill* bores holes up to 2 inches in diameter in rock, concrete, or masonry for the placement of internal charges. The *pavement breaker* is used to shatter the hard surface of roads before drilling boreholes with an auger for cratering charges. The *wood-boring machine* drills boreholes in wood for the placement of internal charges.

41. Blasting Kits

These kits or sets are assemblies of demolition explosive items, accessories, and tools needed for various jobs. They are issued according to tables of equipment.

a. Electric and Nonelectric Kit. The electric and nonelectric demolition equipment set consists of TNT and M5A1 (Composition C-4) demolition blocks and accessories for electric and nonelectric priming and firing (fig. 44). The set is carried in the engineer squad and platoon demolition chest.

(1) *Components issued as basic kit.* The basic kit consists of items listed below. These may be requisitioned separately for replacement purposes.

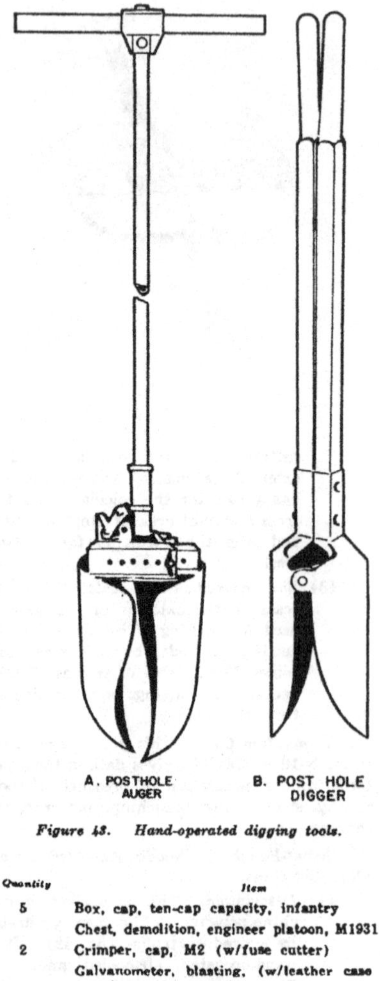

Figure 43. Hand-operated digging tools.

Quantity	*Item*
2	Bag, canvas, carrying, demolition kit
1	Blasting machine, ten-cap capacity
5	Box, cap, ten-cap capacity, infantry
	Chest, demolition, engineer platoon, M1931
2	Crimper, cap, M2 (w/fuse cutter)
	Galvanometer, blasting, (w/leather case and carrying strap)
2	Knife, pocket
2	Pliers, lineman's (w/side cutter), length 8 in

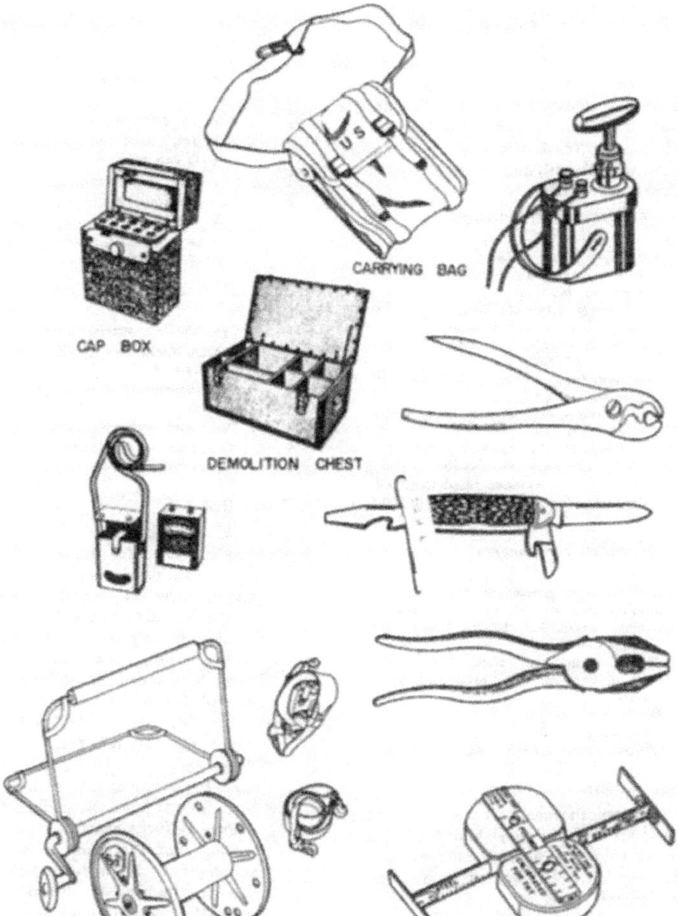

Figure 44. Electric and nonelectric blasting kit.

Quantity	Item
1	Reel, wire, firing, 500 ft, RL-39A, (w/ carrying straps, w/winding device, w/ spool, w/o wire)
2	Tape, computing, demolition charge

(2) *Components issued separately.* The following items are required to complete the kit and should be on hand at all times. These items are not supplied with the kit, and must be requisitioned separately.

(*a*) *Nonexplosive components.*

Quantity	Item
60	Adapter, priming, M1A4
2	Adhesive, paste, for demolition charges, ½-lb can, M1
1	Cable, power, electrical, firing, vinyl polymer insulation, two conductor, No. 18 AWG stranded, 500-ft coil
50	Clip, cord, M1, detonating
6	Insulation tape, electrical, black adhesive, ¾-in wide
1	Sealing compound, blasting cap, waterproof, ½-pt can
2	Twine, hemp, No. 18,8-oz. ball
2	Wire, electrical, annunciator, waxed double cotton wrapped insulation, solid single conductor, No. 20 AWG, 200-ft coil

(*b*) *Explosive components.*

Quantity	Item
50	Cap, blasting, special, electric M6
50	Cap, blasting, special, nonelectric M7
40	Charge, demolition, block, M5A1 2½-lb Comp C-4
50	Charge, demolition, block, 1-lb (TNT)
5	Cord, detonating, fuze, primacord, 100-ft spool
5	Destructor, explosive, universal M10
2	Fuze, blasting, time, 50-ft coils
50	Igniter, blasting fuze M60, weatherproof

b. Nonelectric Kit.

(1) *Components issued as basic kit.* The basic kit (fig. 45) consists of the items listed below. These items may also be requisitioned separately for replacement purposes.

Quantity	Item
2	Bag, canvas, carrying, demolition kit
2	Box, cap, 10-cap capacity, infantry
2	Crimper, cap, M2 (w/fuse cutter)
2	Knife, pocket
2	Tape, computing, demolition charge

(2) *Components issued separately.* The following items are required to complete the kit and should be on hand at all times. They must be requisitioned separately, however.

(*a*) *Nonexplosive components.*

Quantity	Item
20	Adapter, priming, M1A4
2	Adhesive, paste, for demolition charges, ½ lb can M1
50	Clip, cord, M1, detonating
2	Insulation tape, electrical, black adhesive, ¾ in wide
1	Sealing compound, blasting cap, waterproof, ½ pt can

(*b*) *Explosive components.*

Quantity	Item
50	Cap, blasting, special, nonelectric, M7
40	Charge, demolition, block, M5A1, 2½ lb, Comp C-4
2	Cord, detonating, fuze, primacord, 100 ft spool
2	Destructor, explosive, universal, M10
2	Fuze, blasting, time, 50-ft coils
50	Igniter, blasting fuze, M60, weatherproof

c. Earth Rod Kit.

(1) *Use.* This kit (fig. 46), is used for making holes for demolition or constructional purposes as deep as 6 feet and as large as several inches in diameter in earth and soft shale. It is not usable in rock or other hard material. The rod is driven into the earth by the propelling charge, which is exploded in the firing chamber. A removable handle (extractor rod), which fits through the holes in the firing chamber, and an extension are used for gripping and lifting the rod or pulling it from the earth. A linear charge is furnished for enlarging the diameter of the hole. A forked inserting rod also is furnished for inserting improvised linear charges when the standard ones are not available.

(2) *Components.*

Note. The item letters in (*a*) and (*b*) below are keyed to figure 46.

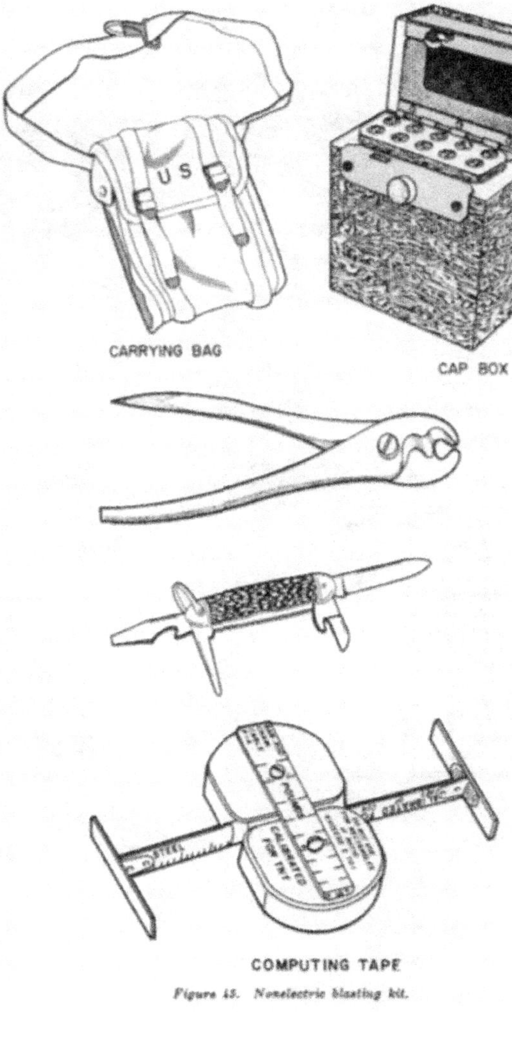

Figure 45. Nonelectric blasting kit.

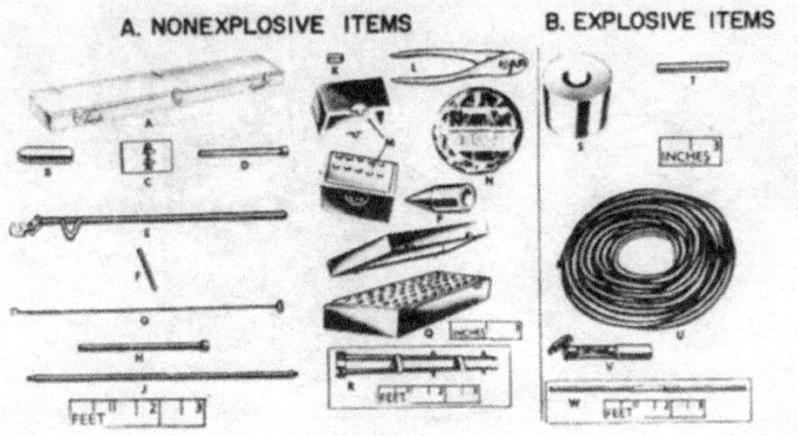

Figure 46. Earth rod kit.

(*a*) *Nonexplosive items.*

Item letter	*Quantity*	*Item*
A	1	Chest
B	1	Chamber, firing
C	1	Plate, base, extractor, assy
D	1	Rod, extension
E	1	Extractor, rod
F	1	Rod, handle and starting
G	1	Rod, inserting
H	2	Rod, intermediate
J	2	Rod, main, long
K	100	Adapter, priming, explosive, M1A3 or M1A4
L	1	Crimper, cap, M2 (w/fuse cutter)
M	1	Box, cap, 10-cap capacity, infantry
N	2	Insulation tape, electrical, black adhesive cotton, ¾-inch wide
P	100	Point
Q	2	Box, cap, 50-cap capacity, engineer
R	1	Tripod

(*b*) *Explosive items.*

Item letter	*Quantity*	*Item*
S	100	Charge, propelling, earth rod, M12 (w/primer, M44)
T	100	Cap, blasting, special, non-electric (type I (J-1 PETN))
U	2	Fuze, blasting, time 50-ft coils
V	200	Igniter, time blasting fuze, M2, weatherproof
W	100	Charge, demolition, linear (two 3-ft sections and one connecting sleeve)

CHAPTER 2

FIRING SYSTEMS

Section I. NONELECTRIC FIRING SYSTEM

42. Introduction

Two types of systems for firing explosives are in general use—electric and nonelectric. Both have their individual priming methods and materials. In addition, detonating cord may be used with both systems to improve and make them more efficient and effective, as described in paragraphs 63 through 71.

43. System Components and Assembly for Detonation

A nonelectric system is an explosive charge prepared for detonation by means of a nonelectric blasting cap. The priming materials consist of a nonelectric blasting cap, which provides the shock adequate to initiate the explosive, and the time fuse, which transmits the flame that fires the blasting cap. The assembly of the nonelectric system follows.

a. Cut and discard a 6-inch length from the free end of the time fuse (A, fig. 47). Do this to be sure that there is no chance of misfire from a damp powder train because of the absorption of moisture from the open air. Then cut off a minimum of 3 feet of time fuse to check the burning rate. A more exact check may be made by marking off 1-foot lengths, timing them separately, and taking the average.

b. Cut the time fuse long enough to permit the person detonating the charge to reach a safe distance by walking at a normal pace before the explosion. This cut should be made squarely across the time fuse.

c. Take one blasting cap from the cap box, inspect it, hold it with the open end down, and shake it gently or bump the hand holding it against the other hand, to remove any dirt or foreign matter. *Never tap the cap with a hard object or against a hard object. Never blow into the cap. Do not insert anything into the cap to remove any dirt or foreign material.*

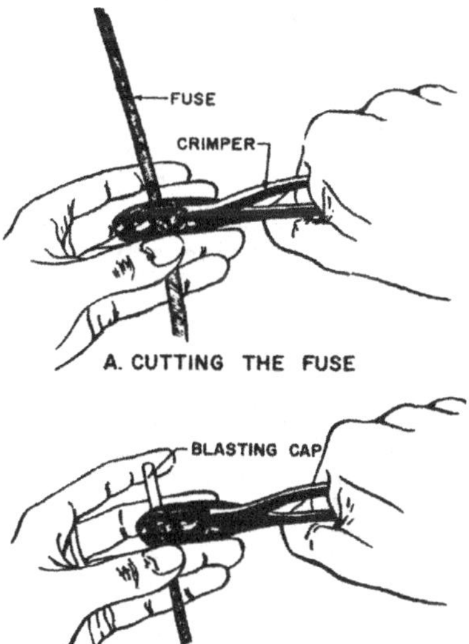

Figure 47. Capping the fuse.

d. Hold the time fuse vertically with the square cut end up and *slip the blasting cap gently down over it so that the flash charge in the cap is in contact with the end of the time fuse; if not, it may misfire. Never force the time fuse into the blasting cap by twisting or any other method.* If the end is flattened or it is too large to enter the blasting cap freely, roll it between the thumb and fingers until the size is reduced to permit free entry.

e. After the blasting cap has been seated, grasp the time fuse between the thumb and third finger of the left hand and extend the forefinger over the end of the cap to hold it firmly against the end of the time fuse. Keep a slight pressure on the closed end of the cap with the forefinger (B, fig. 47).

f. Slide the second finger down the outer edge of the blasting cap to guide the crimpers (B, fig. 47) and thus obtain accurate crimping, even in darkness.

g. Crimp the blasting cap at a point 1/8 to 1/4 of an inch from the open end. *A crimp too near the explosive in the blasting cap may cause detonation. Point the cap out and away from the body during crimping.* Double-crimp the cap, if necessary, for weatherproofing.

Note. If the blasting cap should remain in place several days before firing, protect the joint between the cap and the time fuse with a coating of sealing compound or some similar substance. *(As this sealing compound, a standard issue, does not make a waterproof seal, submerged charges should be fired immediately).*

h. Pass the end of the time fuse through the priming adapter. (The time fuse should move through the adapter easily.) Then pull the cap into the adapter until it stops, insert into the cap well of the explosive, and screw the adapter into place. If no priming adapter is available, insert the capped time fuse into the cap well and tie it in place with a string or fasten it with adhesive tape or some other available material. (For details of nonelectric priming of demolition blocks, see para 44–46.)

i. Attach M60 weatherproof fuse lighter as follows:

(1) Unscrew the fuse holder cap two or three turns but do not remove. Press the shipping plug into the lighter to release the split collet (fig. 40), and rotate the plug as it is removed.

(2) Insert the free end of the time fuse in place of the plug until it rests against the primer.

(3) Tighten the cap sufficiently to hold the fuse in place and thus weatherproof the joint.

(4) To fire, remove the safety pin, hold the barrel in one hand, and pull on the pull ring with the other, taking up the slack before making the final strong pull. In the event of a misfire, the M60 can be reset quickly without disassembly by pushing the plunger all the way in and attempting to fire as before. (It cannot be reset underwater, however, because water can enter the interior of the nylon case through the holes in the pull rod. The fuse lighter is reusable if the primer is replaced.)

j. Light the time fuse with a match by splitting the fuse at the end (fig. 48), placing the head of an unlighted match in the powder train, and then lighting the inserted match head with a flaming match or by rubbing the abrasive on the match box against it.

Note. The M2 weatherproof fuse lighter (fig. 39) may be attached by sliding the fuse retainer over the end of the fuse, firmly seating it, and applying sealing compound at the joint between the time fuse and the lighter to retard any flash that may come from lighting the time fuse. In firing, hold the barrel in one hand and pull on the pull ring with the other.

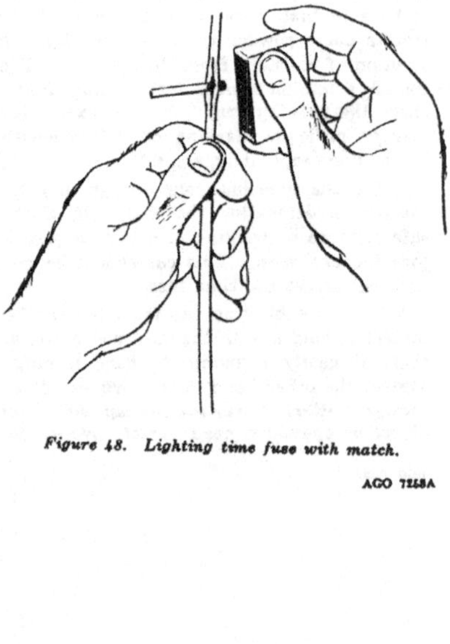

Figure 48. Lighting time fuse with match.

44. Nonelectric Priming of Demolition Blocks

a. With Priming Adapter. Priming adapters simplify the priming of military explosives with threaded cap wells. The shoulder inside one end of the adapter is large enough to admit time fuse or detonating cord, but too small for a blasting cap. The other end of the adapter fits the internal thread of threaded cap wells in military explosives. The nonelectric priming components are assembled as shown in figure 49.

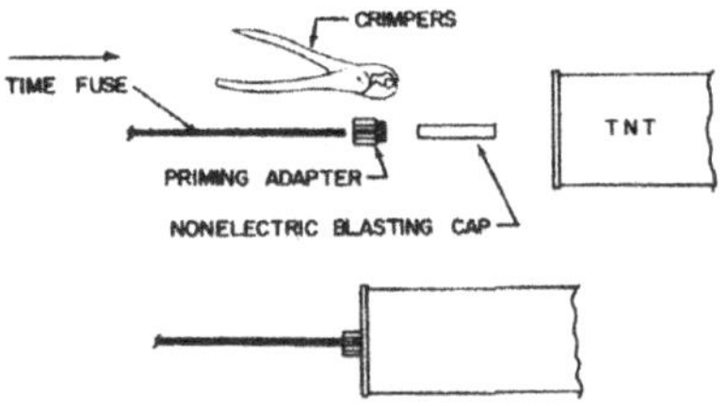

Figure 49. Nonelectric priming, using priming adapter.

b. Without Priming Adapter. When a priming adapter is not available but explosive blocks have threaded cap wells, they are primed as follows (see method 1, fig. 50):

(1) Wrap a string tightly around the block and tie it securely leaving about 6 inches of loose string on each end after making the tie (method 1, fig. 50).

(2) Insert a blasting cap with fuse attached into the cap well.

(3) Tie the loose string around the fuse to prevent the blasting cap from being separated from the block.

c. Without Cap Well.

(1) If the demolition block has no cap well, make a hole in the end large enough to receive the cap with a pointed instrument or the pointed leg of the crimper handle (method 2, fig. 50).

(2) Insert fused cap into hole, grasp fuse with thumb and forefinger at top of hole, and remove fused cap from the block of explosive.

(3) Using string approximately 40 inches in length, tie two half hitches around fuse so the tie will be at the top of the hole when reinserted.

(4) Insert fused cap into hole and wrap long end of string around the block of explosive a minimum of three times along the long axis, each time changing the direction of tie with a half turn around the time fuse, keeping the string taut.

(5) Tie off around the time fuse at top of hole with two half hitches.

Note. Never try to force a cap into an expedient fuse well that is too small to admit it easily.

45. Nonelectric Priming of M1 Chain Demolition Block

The M1 chain, demolition block is primed nonelectrically by fastening a nonelectric blasting cap at a point at least 6 inches in from one of the free ends of the detonating chord chain as shown in figure 51. *The explosive end of the cap should point toward the demolition blocks.* The firing of the blasting cap detonates the cord, which in turn detonates the explosive blocks.

46. Nonelectric Priming of Plastic Explosive (C3 and C4)

a. The M5 (C3) and M5A1 (C4) demolition blocks with threaded cap wells or recesses and with or without priming adapters and the M3 (C3) demolition block without a threaded cap well are primed nonelectrically as described in paragraph 44 and shown in figures 49 and 50.

b. Plastic explosive removed from the container is primed nonelectrically by molding it around a fused blasting cap (fig. 52). The explosive must be at least one inch thick at the explosive end of the blasting cap and ½ inch thick at the sides to insure detonation.

47. Nonelectric Priming of Dynamite

Dynamite cartridges may be primed nonelectrically at either end or at the side. End prim-

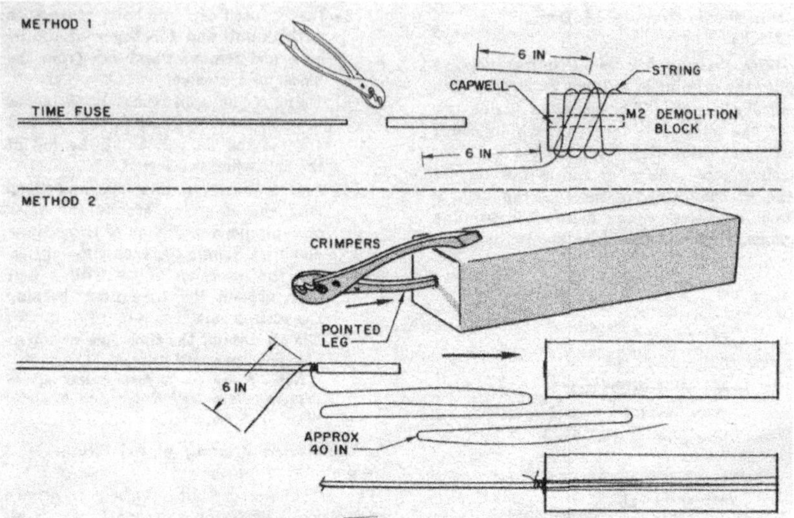

Figure 50. Nonelectric priming without priming adapter.

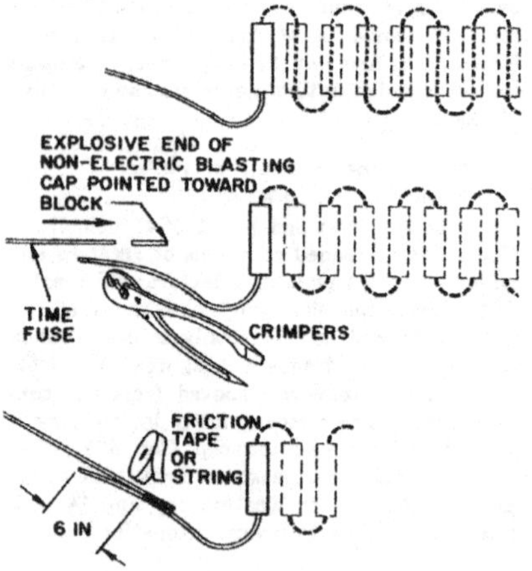

Figure 51. Nonelectric priming of M1 chain demolition block.

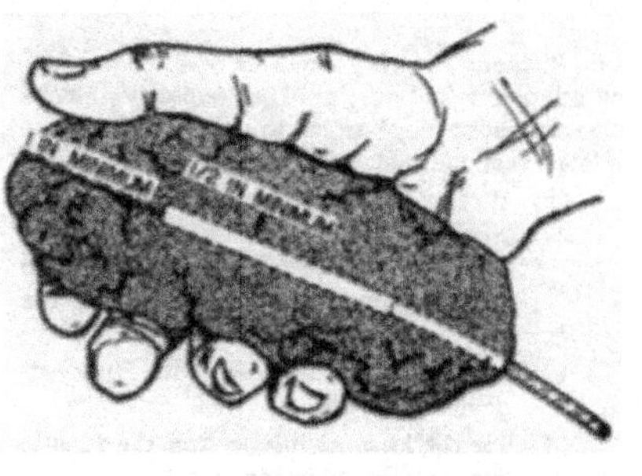

Figure 52. Nonelectric priming of molded plastic explosive.

ing is used either when a whole case is fired or when placed charges require no tamping.

a. End Priming Method.

(1) Punch a hole in the end of the cartridge (fig. 53).

(2) Insert a fused blasting cap.

(3) Tie the cap and fuse securely in the cartridge.

b. Weatherproof Priming Method.

(1) Unfold the wrapping at the folded end of the cartridge.

(2) Punch a hole in the exposed dynamite.

(3) Insert a fused blasting cap into the hole.

c. Side Priming Method (for Charges in Tamped Boreholes).

(1) Punch a hole in the cartridge about 1½ inches from one end (fig. 54).

(2) Point the hole so that the blasting cap when inserted, will be nearly parallel with the side of the cartridge and the

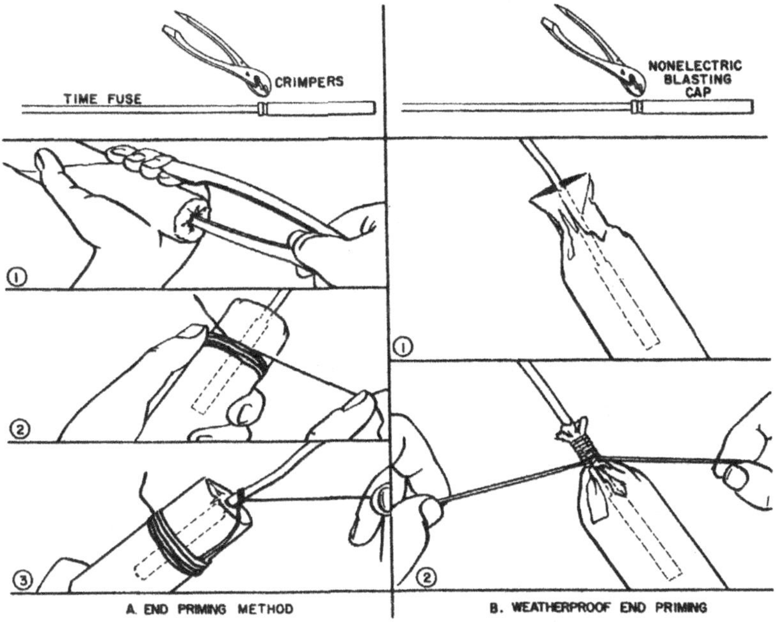

Figure 53. Nonelectric priming of dynamite at end.

(4) Close the wrapping.

(5) Fasten the cap and fuse securely with a string or length of tape (fig. 53).

(6) Apply weatherproofing compound to tie.

explosive end of the cap will be at a point at about half the length of the cartridge.

(3) Insert a fused blasting cap into the hole.

(4) Wrap a string tightly around the fuse and then around the cartridge, making two or three turns before tying the string (fig. 54).

(5) Moistureproof the primer by wrapping a string closely around the cartridge, extending it an inch on each side of the hole to cover it completely.

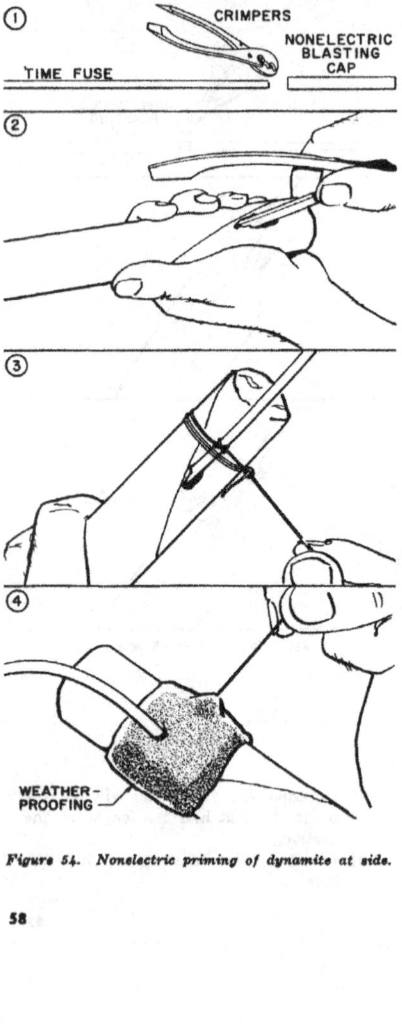

Figure 54. Nonelectric priming of dynamite at side.

Then cover the string with a water-repellent substance.

48. Nonelectric Priming of Ammonium Nitrate and Nitramon Charges

The ammonium nitrate and nitramon charges are primed nonelectrically, as follows:

a. Place a fused blasting cap in the cap well on the side of the container (fig. 55).

b. Tie a string around the fuse and then around the cleat above the blasting cap.

Note. A primed block of TNT placed on top of the charge is recommended to insure positive detonation.

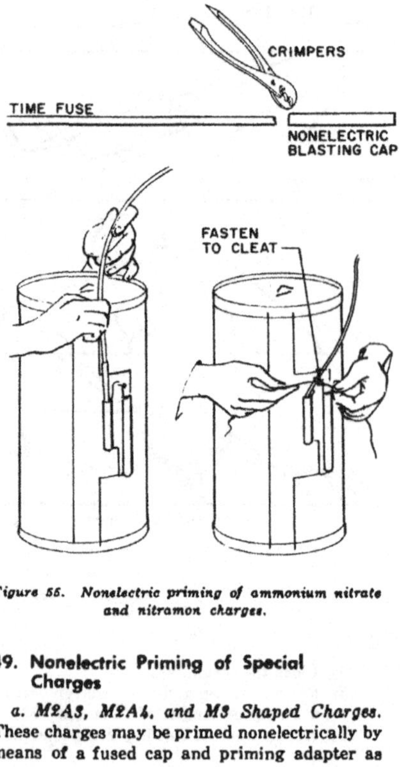

Figure 55. Nonelectric priming of ammonium nitrate and nitramon charges.

49. Nonelectric Priming of Special Charges

a. M2A3, M2A4, and M3 Shaped Charges. These charges may be primed nonelectrically by means of a fused cap and priming adapter as

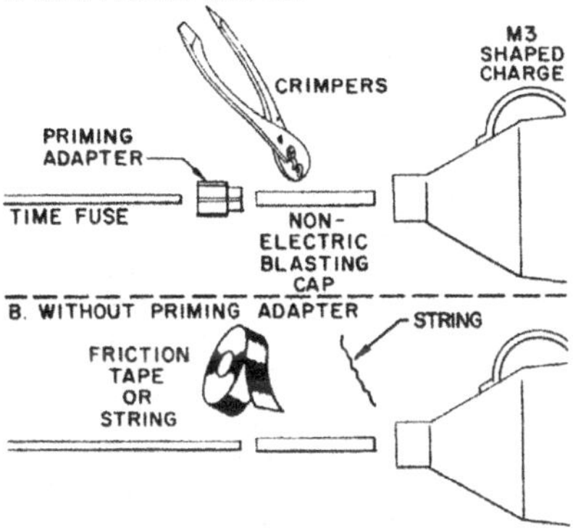

Figure 56. Nonelectric priming of shaped charges.

shown in figure 56. If a priming adapter is not available, the primer may be held in the cap well by a string or friction tape.

b. Bangalore Torpedo. The bangalore torpedo may be primed by assembling a length of time fuse and a nonelectric blasting cap in a priming adapter and screwing the assembly into the cap well of a torpedo section (fig. 57). A section may also be primed nonelectrically by a pull type firing device, with a nonelectric blasting cap crimped on the base, screwed into the cap well (fig. 57).

c. M118 Demolition Charge. This charge is commonly known as sheet explosive. It is packaged in four sheets ¼ × 3 × 12 inches, with an adhesive on one side. It is primed by three methods (fig. 58):

(1) By inserting and holding a nonelectric blasting cap in a grove or notch cut in the charge or

(2) By placing the cap between two pieces of explosive or

(3) By placing the cap between an overlap in the charge.

50. Nonelectric Misfires

a. Prevention. Working on or near a misfire is the most hazardous of all blasting operations. A misfire should be extremely rare if these procedures are followed closely:

(1) Prepare all primers properly.

(2) Load charges carefully.

(3) Place primer properly.

(4) Perform any tamping operation with care to avoid damage to an otherwise carefully prepared charge.

(5) Fire the charge according to the proper technique.

(6) If possible, use dual firing systems

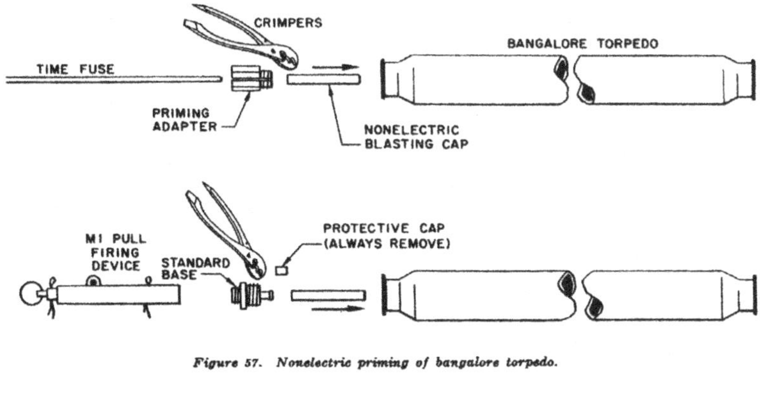

Figure 57. Nonelectric priming of bangalore torpedo.

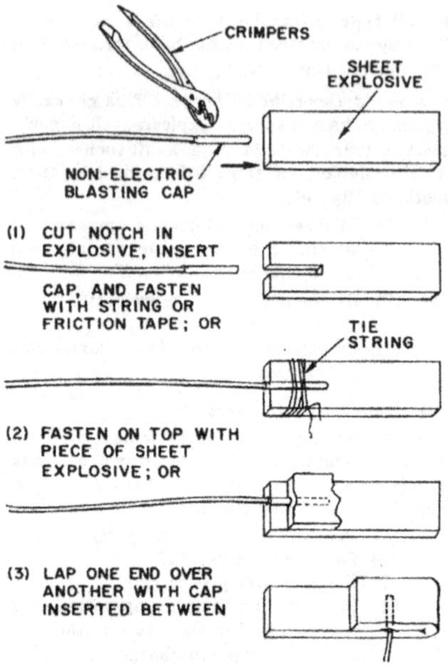

Figure 58. Nonelectric priming of sheet explosive.

(para 72–75). If both systems are properly assembled, the possibility of a misfire is reduced to a minimum.

b. The Handling of Nonelectric Misfires. Occasionally, despite all painstaking efforts, a nonelectric misfire will occur. Investigation and correction should be undertaken only by the man that placed the charge. For a charge primed with a nonelectric cap and time fuse, the procedure is as follows:

(1) Delay the investigation of the misfire at least 30 minutes after the expected time of detonation. This should be ample time for any delayed explosion to take place because of a defective powder train in the fuse. Under certain combat conditions, however, immediate investigation may be necessary.

(2) If the misfired charge is not tamped, lay a new primer at the side of the charge, without moving or disturbing it, and fire.

(3) If the misfired charge has no more than a foot of tamping, attempt to explode it by detonating a new 2-pound primer placed on top.

(4) If the misfired charge is located in a tamped borehole, or if the tamped charge is so situated as to make method (3) above impractical, remove the tamping by means of wooden or nonmetallic tools. Avoid accidentally digging into the charge. Also, the tamping may be blown out by means of a stream of compressed air or water if either is available. Constant checking of the depth of the borehole from the ground surface or the top of the charge during digging will minimize the danger of striking the charge. When the charge has been uncovered within 1 foot, insert and detonate a new 2-pound primer.

(5) An alternate method of reaching a deep misfired charge is to drill a new hole within one foot of the old one and to the same depth. A 2-pound primed charge is then placed in the new hole to detonate the misfired charge. Extreme care is required in drilling the new hole to avoid striking the old misfired charge or placing the new charge too far away to induce detonation.

SECTION II. ELECTRIC FIRING SYSTEM

51. Components and Assembly for Detonation

The electric firing system provides the electric spark or impulse to initiate detonation. The electric impulse travels from the power source through the lead wires to fire the cap.

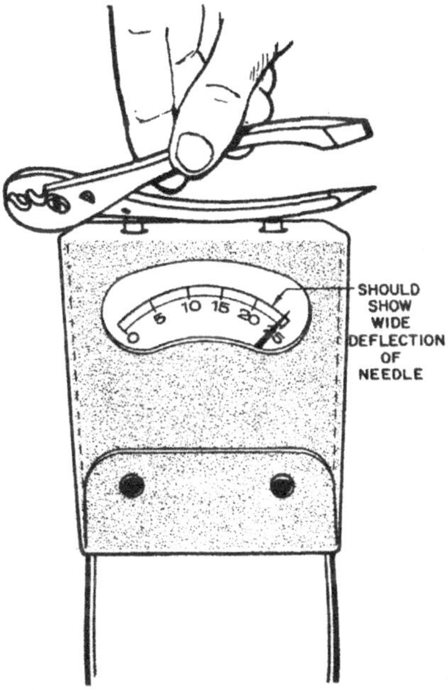

Figure 59. Testing the galvanometer.

The chief components of the system are the blasting cap, firing wire and reel, and the blasting machine. The preparation of the explosive charge for detonation by electrical means is called electric priming. Priming methods are described below.

a. Testing the Cap.

(1) Test the galvanometer (para 36) (fig. 59).

(2) Remove the short circuit shunt (prevents accidential firing) from the lead wires of the electric blasting cap (fig. 60).

(3) Touch one cap lead wire to one galvonmeter post and the other cap lead wire to the other. If the instrument registers a flowing current, the blasting cap is satisfactory; if not, the cap is defective and should not be used. During the test, *always point the explosive end of the blasting cap away from the body.*

b. Placing the Cap in the Explosive.

(1) Pass the lead wires through the slot of the adapter and pull the cap into place in the adapter.

(2) Insert the cap into the cap well of the explosive and screw the adapter into place.

(3) If a priming adapter is not available, insert the electric cap into the cap

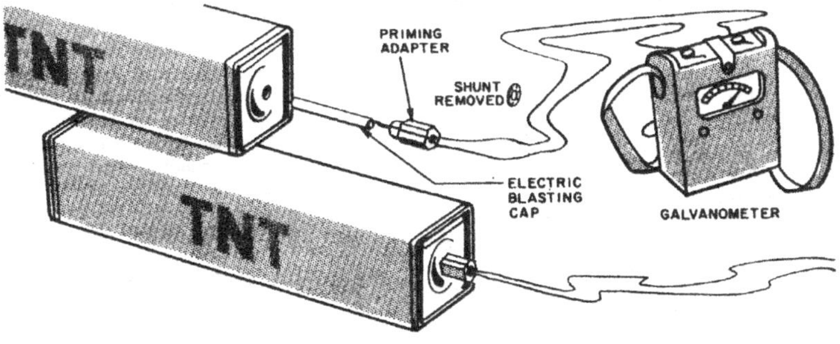

Figure 60. Assembling the electric primer.

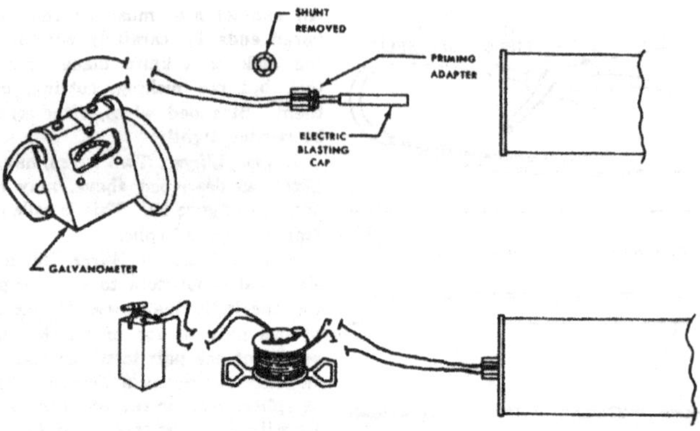

Figure 64. Electric priming of demolition block with priming adapter.

(2) *Without priming adapter.* If a priming adapter is not available and the demolition block has a threaded cap well, insert the electric cap into the cap well and tie the lead wires around the block by two half-hitches or a girth hitch. Leave a small portion of slack wire between the blasting cap and the tie to prevent any pull on the blasting cap (fig. 61).

b. Blocks Without Cap Wells. If the demolition blocks have no threaded cap wells proceed as follows:

(1) Make a hole in the end *large enough to receive the cap* with a pointed nonspark instrument or the pointed leg of the crimper handle (fig. 65).

(2) Insert the cap in this cap well and tie the cap wires around the block by two half hitches or a girth hitch. To prevent pull on the cap, always leave a small portion of slack wire between the blasting cap and the tie.

Note. Never try to force a blasting cap into an expedient fuse well that is too small to admit it easily.

54. Electric Priming of M1 Chain Demolition Block

The M1 chain demolition block is primed elec-

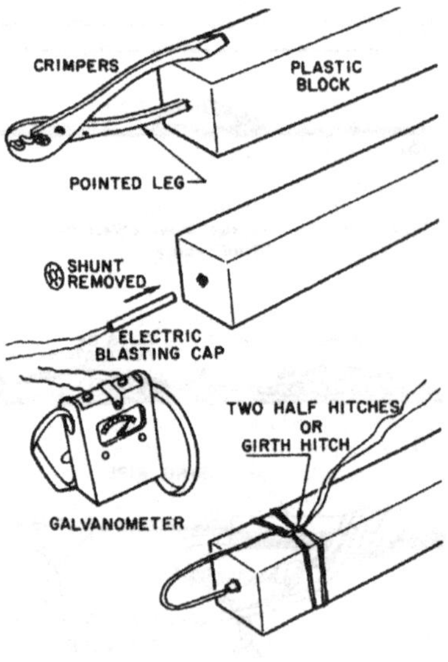

Figure 65. Electric priming of plastic explosive without cap well.

trically by fastening an electric blasting cap to one of the free ends of the detonating cord chain with friction tape or by some other method. *The explosive end of the cap should point toward the chain demolition blocks.* The firing of the blasting cap detonates the cord, which in turn detonates the entire chain.

55. Electric Priming of Plastic Explosives (C3 and C4)

a. The M5 (C3) and M5A1 (C4) demolition blocks with threaded cap wells or recesses and with and without priming adapters and the M3 (C3) demolition block without a threaded cap well are primed electrically as shown in paragraph 53 and figures 61, 64, and 65.

b. Plastic explosive removed from the container is primed electrically by molding it around the blasting cap (fig. 66). The explosive must extend at least 1 inch at the explosive end of the blasting cap and ½ inch at the sides to insure detonation.

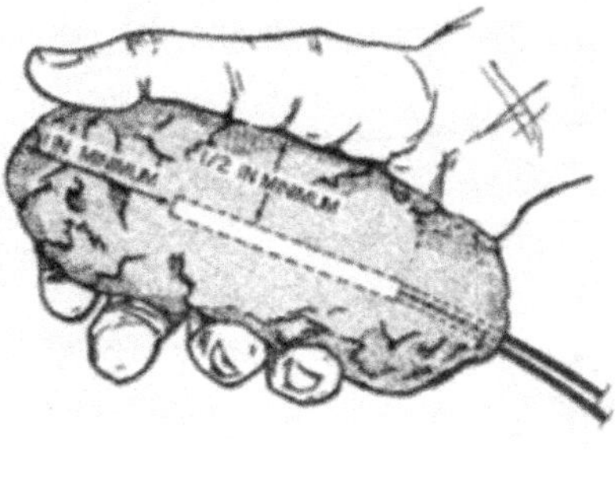

Figure 66. Electric priming of molded plastic explosive.

56. Electric Priming of Dynamite

Dynamite cartridges may be primed electrically at either end or at the side. End priming is used either when a whole case is fired or when placed charges require no tamping.

a. End Priming Method.

(1) Punch a hole in the end of the cartridge (fig. 67).

(2) Insert an electric blasting cap.

(3) Tie the lead wires around the cartridge with two half hitches.

b. Side Priming Method.

(1) Punch a hole in the cartridge about 1½ inches from one end (fig. 68).

(2) Point the hole so that the electric blasting cap, when inserted, will be nearly parallel with the side of the cartridge and the explosive end of the cap will be at a point at about half the length of the cartridge.

(3) Insert the blasting cap into the hole.

(4) Tie the lead wires around the cartridge with two half hitches or fasten with a string or tape.

57. Electric Priming of Ammonium Nitrate and Nitramon Charges

The electric blasting cap is placed in the cap well and the lead wires are looped around the cleat (fig. 69).

58. Electric Priming of Special Charges

a. M2A3, M2A4, and M3 Shaped Charges. These may be primed electrically by means of an electric blasting cap and a priming adapter as shown in figure 70. If a priming adapter is not available, the electric cap may be held in place by a length of string or friction tape.

b. Bangalore Torpedo. The bangalore torpedo may be primed electrically by assembling a blasting cap and priming adapter and screwing the assembly into the cap well of a torpedo section (fig. 71).

c. M118 Demolition Charge. This charge, commonly known as sheet explosive (fig. 72), is packaged in four sheets ¼ x 3 x 12 inches, with an adhesive on one side. It is primed by—

(1) Inserting and holding an electric blasting cap in a groove or notch cut in the charge, or

(2) Placing the cap between two sheets or pieces of explosive or,

(3) Placing the cap between an overlap of the charge.

59. Series Circuits

a. Common Series. This is used for connecting two or more charges fired electrically by a blasting machine (fig. 73). A common series circuit is prepared by connecting one

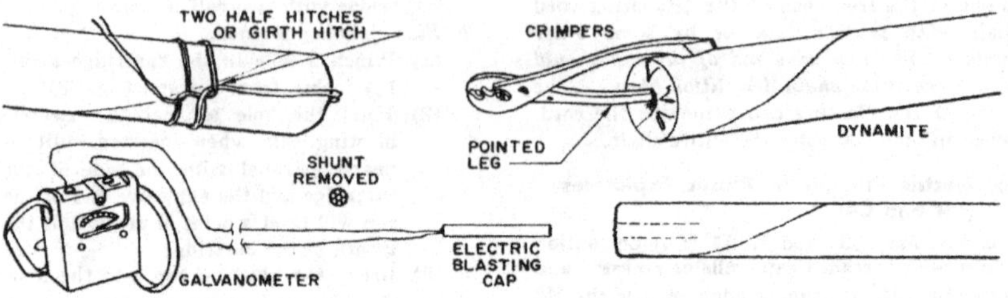

Figure 67. Electric priming of dynamite at end.

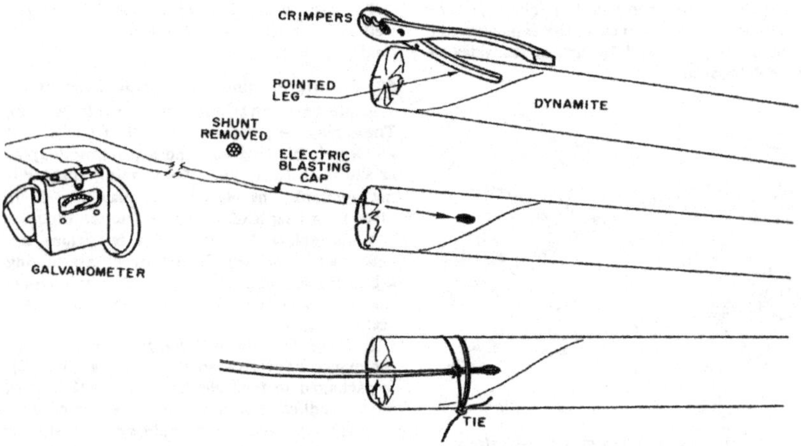

Figure 68. Electric priming of dynamite at side.

blasting cap lead wire from the first charge to one lead wire in the second charge and so on until only two end wires are free, then connecting the free ends of the cap lead wires to the ends of the firing wire. Connecting wires (usually annunicator wire) are used when the distance between blasting caps is greater than the length of the usual cap lead wires.

b. "Leapfrog" Series. The "leapfrog" method of connecting caps in series (fig. 73) is useful for firing ditching charges or any long line of charges. It consists of omitting alternate

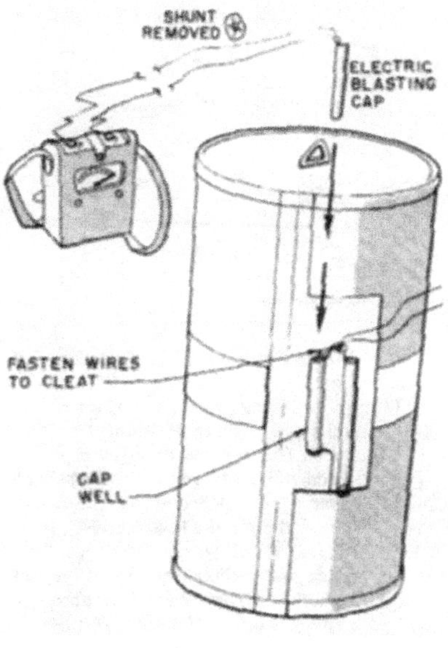

Figure 69. Electric priming of ammonium nitrate and nitramon charges.

charges on the way and then connecting them to form a return path for the electric impulse to reach the other lead of the firing wire. This brings both end wires out at the same end of the line of charges, and thus eliminates laying a long return lead from the far end of the line of charges back to the firing wire.

60. Testing Electric Wires and Circuits

a. Firing Wires May be Tested as Follows:

(1) Check galvanometer by holding a piece of metal across its terminals (para 36; fig. 59).

(2) Separate the firing wire at both ends, and touch those at one end to the galvanometer posts. The needle should not move. If it does, the firing wire has a short circuit (fig. 74); or

(3) Twist the wires together at one end and touch those at the other to the galvanometer posts. This should cause a wide deflection in the needle. (See note at end of *c* (2), below). No movement of the needle indicates a break; a slight movement, a point of high resistance.

Note. Firing wire may be tested on the reel, *but should be tested again after unreeling, which may separate broken wires unnoticed when reeled.*

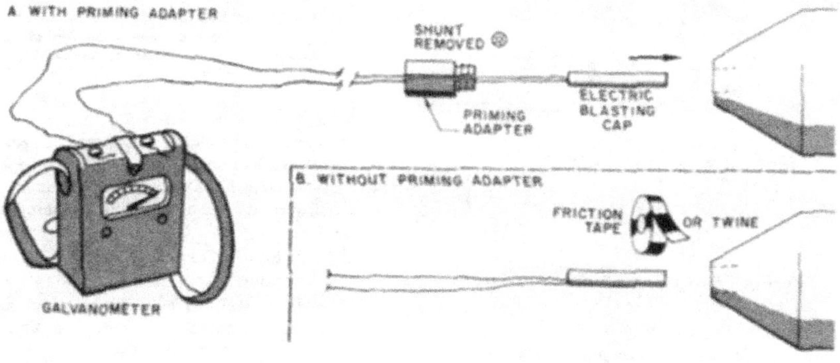

Figure 70. Electric priming of shaped charges.

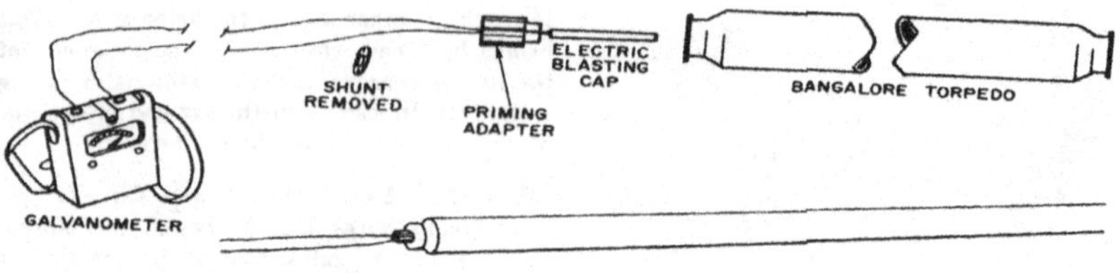

Figure 71. Electric priming of bangalore torpedo.

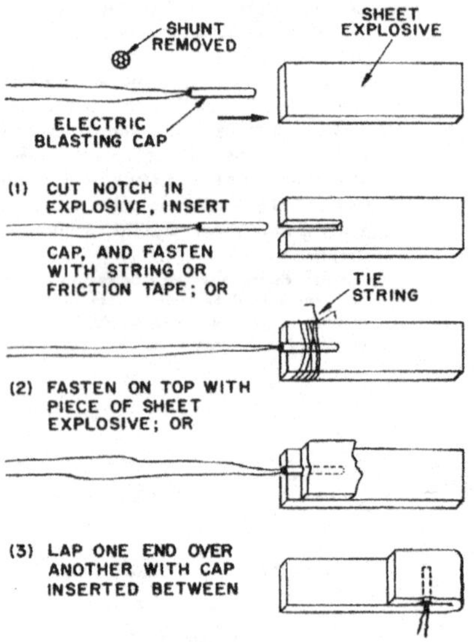

Figure 72. Electric priming of sheet explosive.

b. Series Circuits May be Tested as Follows:

(1) Connect charges as shown in figure 75.

(2) Touch the free ends of the lead wires to the galvanometer posts. This should cause a wide deflection of the needle.

c. The Entire Circuit May be Tested as Follows:

(1) Splice firing wires to series circuit and move to firing position.

(2) Touch free ends of firing wire to galvanometer posts (fig. 76). This should cause a wide deflection of the needle. If not, the circuit is defective. If defective, shunt wires. Then go down range and recheck the circuit, repeating *a*(2), (3) and *b*(1), (2) above. If a splice is found defective, resplice the wires. If a cap is found defective, replace it. Then test the entire circuit again to make sure that all breaks have been located before attempting to fire the charge.

Note. To get a "wide deflection of the needle" the galvanometer battery should be in good condition.

61. Electric Misfires

a. Prevention of Electric Misfires. In order to prevent misfires, make one demolition specialist responsible for all electrical wiring in a demolition circuit. He should do all splicing to be sure that—

(1) All blasting caps are included in the firing circuit.

(2) All connections between blasting cap wires, connecting wires, and firing wires are properly made.

(3) Short circuits are avoided.

(4) Grounds are avoided.

(5) The number of blasting caps in any

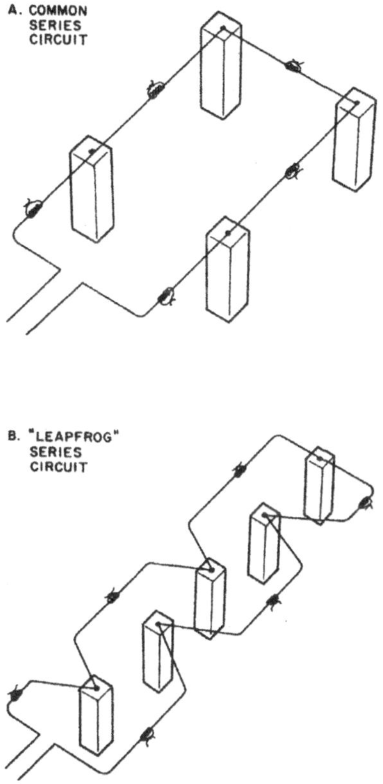

Figure 73. Series circuits.

circuit does not exceed the rated capacity of the power source on hand.

b. Cause of Electric Misfires. Common specific causes of electric misfires include—

(1) Inoperative or weak blasting machine or power source.

(2) Improperly-operated blasting machine or power source.

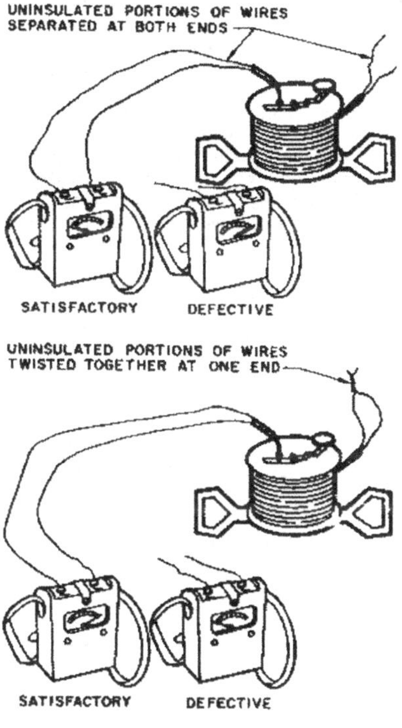

Figure 74. Testing firing wire.

(3) Defective and damaged connections, causing either a short circuit, a break in the circuit, or high resistance with resulting low current.

(4) Faulty blasting cap.

(5) The use in the *same circuit* of old type (J–2) special blasting caps made by different manufacturers.

(6) The use of more blasting caps than the power source rating permits.

c. Handling Electric Misfires. Because of the hazards of burning charges and delayed explosions, electric misfires must be handled with extreme caution. A burning charge may occur with the use of electric as well as nonelectric caps. Misfires of charges primed with detonating cord fired by electric blasting caps are handled as described in paragraph 71. If the charge is dual-primed electrically and below ground, wait 30 minutes before investigating to make sure that the charge is not burning; or if dual-primed above ground, wait 30 minutes before investigating. On the other hand, if the electric misfire is above ground and the charge is not dual-primed, investigate immediately. If the system is below ground and not dual-primed, proceed as follows—

(1) Check the firing wire connection to the blasting machine or power source terminals to be sure that the contacts are good.
(2) Make two or three more attempts to fire the circuits.
(3) Attempt to fire again, using another blasting machine or power source.
(4) Disconnect the blasting machine firing

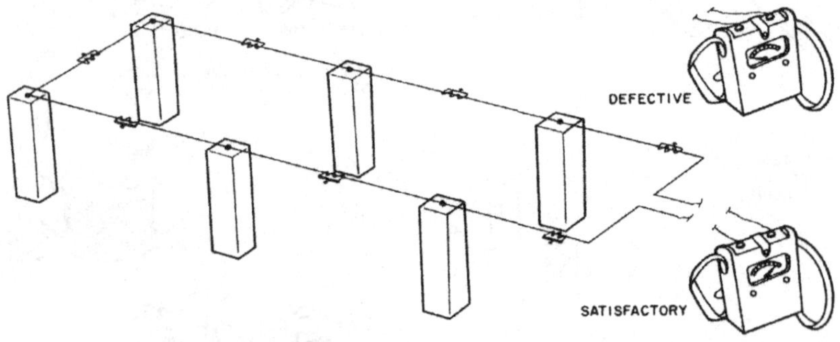

Figure 75. Testing a series circuit.

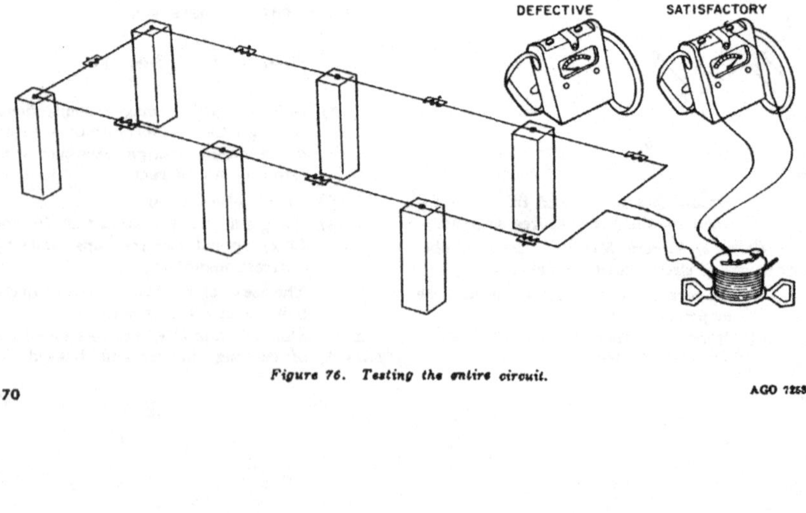

Figure 76. Testing the entire circuit.

wire and wait 30 minutes before further investigation. Before moving on to the charge site, be sure that the firing wires at the power source end of the circuit are shunted to avoid any possible static electric detonation.

(5) Check the entire circuit, including the firing wire, for breaks and short circuits.

(6) If the fault is not above ground, remove the tamping material very carefully from the borehole to avoid striking the electric blasting cap.

(7) Make no attempt to remove either the primer or the charge.

(8) If the fault is not located by the removal of the tamping material to within 1 foot of the charge, place a new electric primer and 2 pounds of explosive at this point.

(9) Disconnect the blasting cap wires of the original primer from the circuit.

(10) Connect the wires of the new primer in their place.

(11) Replace the tamping material.

(12) Initiate detonation. Detonation of the new primer will fire the original charge.

Note. In some cases it may be more desirable or expedient to drill a new hole within a foot of the old one at the same depth to avoid accidental detonation of the old charge and then place and prime a new 2-pound charge.

62. Premature Detonation by Induced Currents and Lightning

a. Induced Currents. The premature detonation of electric blasting caps by induced radio frequency (RF) current is possible. Table VII, showing the minimum safe distance *versus* transmitter power, indicates the distance beyond which it is safe to conduct electrical blasting even under the most adverse conditions. Mobile type transmitters are prohibited within 150 feet of any electric blasting caps or electrical blasting system. If blasting distances are less than those shown in table VII, the only safe procedure is to use a nonelectric system, which cannot be prematurely detonated by RF currents. If, however, the use of the electric system is necessary, follow precautions given in AR 385–63.

Table VII. Minimum Safe Distance for RF Transmitters

Fixed transmitters	
Transmitter power (watts)	Minimum distance (ft.)
5–25	100
25–50	150
50–100	220
100–250	350
250–500	450
500–1,000	650
1,000–2,500	1000
2,500–5,000	1500
5,000–10,000	2200
10,000–25,000	3500
25,000–50,000	5000
50,000–100,000	7000

b. Lightning. Lightning is a hazard to both electric and nonelectric blasting charges. A strike or a nearby miss is almost certain to initiate either type of circuit. Lightning strikes, even at remote locations, may cause extremely high local earth currents and shock waves that may initiate electrical firing circuits. The effects of remote lightning strikes are multiplied by proximity to conducting elements, such as those found in buildings, fences, railroads, bridges, streams, and underground cables or conduit. Thus, the only safe procedure is to suspend all blasting activities during electrical storms.

Section III. DETONATING CORD PRIMING

63. Components

a. Of all primers for explosive charges, detonating cord is probably the most versatile and in many cases the most easily installed. It is especially applicable for underwater and underground blasting, as the blasting cap of the initiating system may remain above the water or ground.

b. The detonating cord primer consists, generally, of a length of detonating cord and the means of detonation which may be an electric blasting cap initiated by a blasting machine or power source or a nonelectric blasting cap initiated by a fuse lighter and a length of time fuse. The blasting cap of either the electric or nonelectric system is attached to the free end of the detonating cord by means of a length of string, wire, or friction tape. Detonating cord primers are usually tied around the explosive block; however, in situations where a close contact between the explosive block and the target is required, a nonelectric blasting cap is crimped on the end of the detonating cord and placed in the cap well.

64. Priming Demolition Blocks

a. *Common Method* (A, fig. 77)

(1) Lay one end of a 4-foot length of detonating cord at an angle across the explosive block as shown in A, figure 77.

(2) Wrap the running end three times over the end laid at an angle and around the block, and on the fourth turn, slip the running end under the three wraps parallel with the other end and then draw tight.

(3) Attach an electric or nonelectric firing system.

b. *Alternate Method No. 1* (*B, fig. 77*).

(1) Tie the detonating cord around the explosive block (on top of the booster, if present) with a clove hitch with two extra turns as shown in B, figure 77. The cord must fit snugly against the blocks and the loops be pushed close together.

(2) Attach an electric or nonelectric firing system.

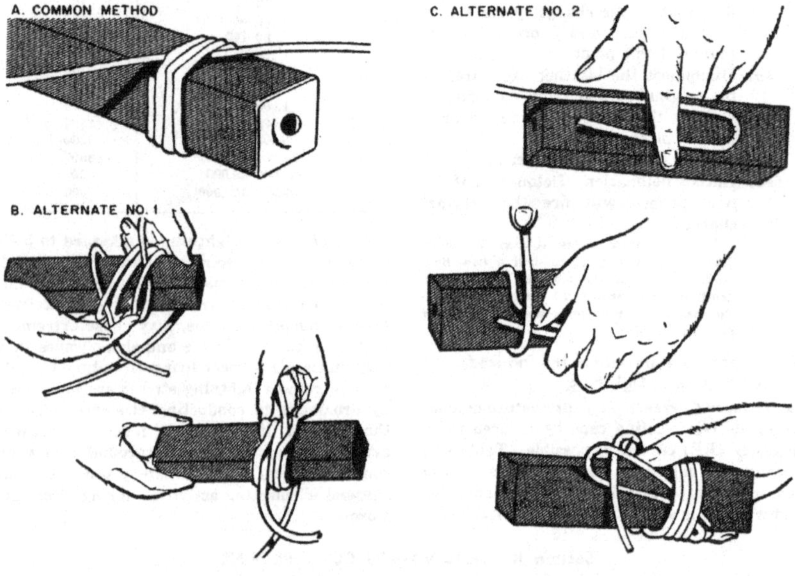

Figure 77. Detonating cord priming of demolition blocks.

c. *Alternate Method No. 2 (C, fig. 77)*

(1) Place a loop of detonating cord on the explosive block as in C, figure 77.

(2) Wrap the detonating cord four times around the block and finally draw the running end through the loop.

(3) Pull until tight.

(4) Attach an electric or nonelectric firing system.

Note. This alternate method is more applicable to *short* than to long detonating cord branch lines or primers.

65. Priming M1 Chain Demolition Block

The M1 chain demolition block has detonating cord running lengthwise through the individual blocks. If an additional length of detonating cord is required, it is connected to the detonating cord of the chain with a clip or square knot. The additional length of detonating cord may be initiated by means of an electric blasting cap and firing device or an nonelectric blasting cap, length of time fuse, and a fuse lighter. If the cord running through the blocks is cut too closely to the end block to permit such a connection, the additional length of detonating cord may be fastened by a clove hitch with two extra turns near the end of the block over the booster (fig. 78).

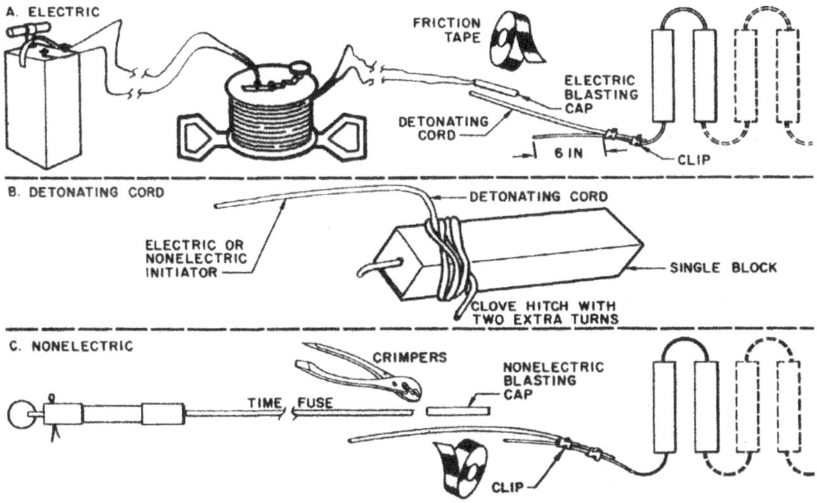

Figure 78. Detonating cord priming of M1 chain demolition block.

66. Priming Dynamite

For use chiefly in boreholes, ditching, or removal of stumps, dynamite is primed by lacing the detonating cord through it. This is done by punching three or four equally-spaced holes through the dynamite cartridge, running the detonating cord back and forth through them, and securing it with a knot or by lacing as shown in figure 79.

67. Priming Plastic Explosive (C3 and C4)

Compositions C3 and C4 when removed from the package are primed with detonating cord, as follows:

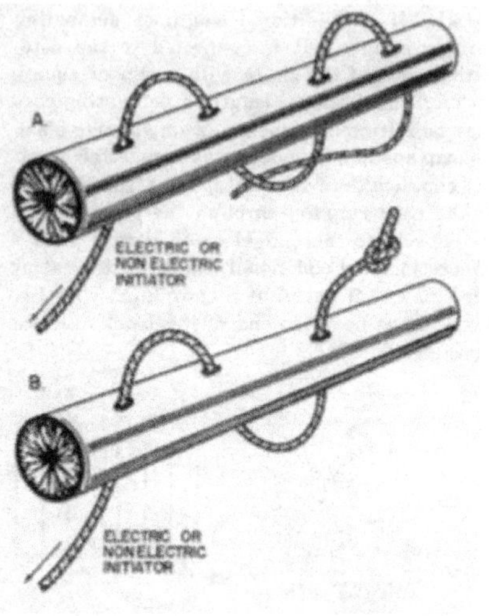

Figure 79. Detonating cord priming of dynamite.

a. Take a 10-inch bight at the end of the detonating cord and tie an overhand knot (fig. 80).

b. Mold the explosive around the knot, leaving at least ½ inch of explosive on all sides and at least 1 inch on each end.

Note. Another method is to cut the block longitudinally, then insert the knot, and fasten with tape or string as shown in figure 80.

68. Priming Ammonium Nitrate and Nitramon Charges

To prime ammonium nitrate and nitramon cratering charges with detonating cord:

a. Pass the detonating cord through the tunnel provided on the one side of the can (fig. 81).

b. Tie an overhand knot on the portion passed through about 6 inches from the end.

c. Attach an electric or nonelectric firing system (fig. 81).

d. For dual priming any firing system may be used for the additional charge—a 1-pound block of TNT—to insure detonation (para 21*c*).

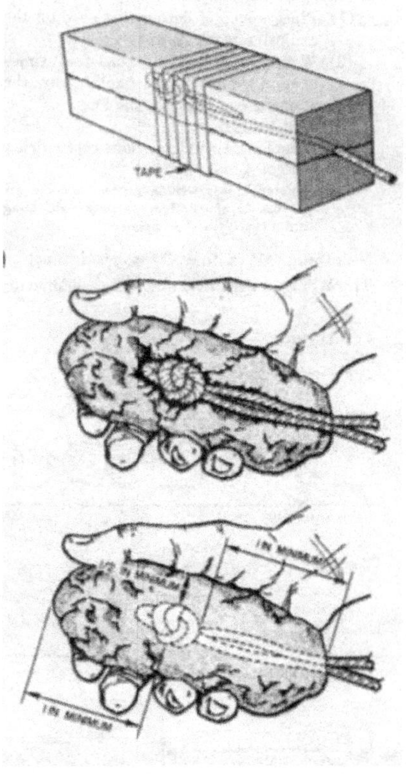

Figure 80. Detonating cord priming of plastic explosive.

69. Priming Special Charges

a. Shaped Charges. M2A3 and M3 shaped charges are primed with a length of detonating cord with a nonelectric blasting cap crimped on (fig. 82), and detonated with an electric or nonelectric firing system.

b. Bangalore Torpedo. A bangalore torpedo may be primed by attaching a length of detonating cord by wrapping six turns directly over

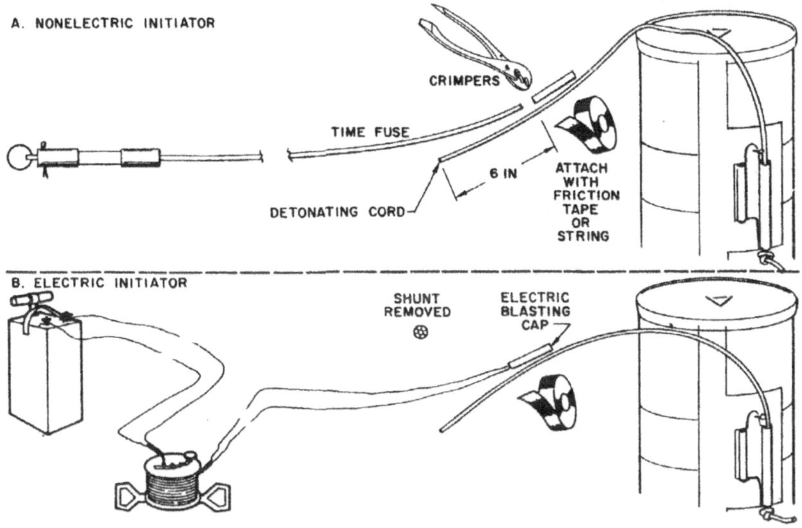

Figure 81. Detonating cord priming of ammonium nitrate and nitramon charges.

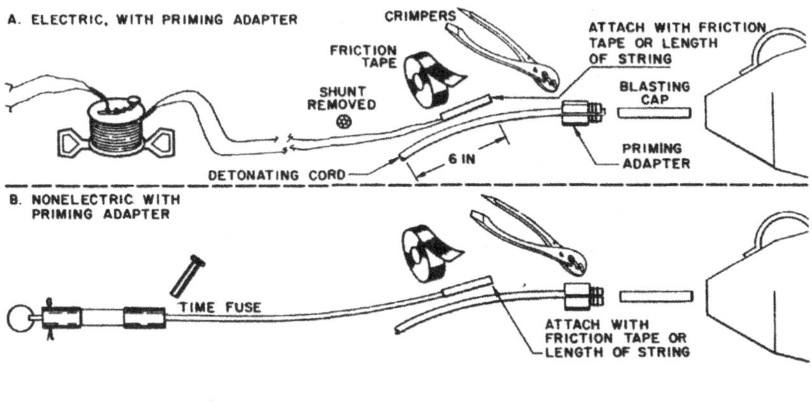

Figure 82. Detonating cord priming of shaped charge.

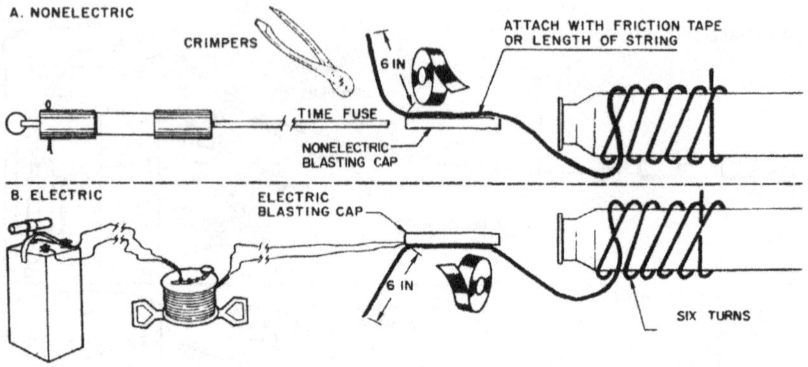

Figure 83. Detonating cord priming of bangalore torpedo.

the booster (fig. 83) and detonating with an electric or nonelectric firing system.

c. Pole Charges. Detonating cord assemblies are excellent primers for pole charges, as the detonating cord spans the distance from the charge to a position where the electric or nonelectric firing system is accessible to the blaster. The method of priming with a dual electric primer is shown in figure 84.

70. Assemblies and Connections

a. Detonating Assemblies.

(1) *Nonelectric.* This detonating assembly consists of a length of detonating cord (approximately 2 feet), a nonelectric blasting cap, a length of time fuse, and a fuse lighter. The blasting cap is crimped to the time fuse and then fastened to the detonating cord (A, fig. 85). The fuse lighter is then fastened to the time fuse. The length of time fuse depends on the time required for the blaster to reach safety after lighting the fuse.

(2) *Electric.* The electric detonating assembly is a length of detonating cord (approximately 2 feet) with an electric blasting cap attached (B, fig. 85).

(3) *Attachment of assembly to system.* The free end of the detonating cord is fastened to the main line by a clip or a square knot as shown in figure 85.

(4) *Advantages.* Many advantages may be gained by th euse of these primers —they may be made up in advance, thus saving time at the target, which is a great advantage when time is a critical factor; and they permit the person in charge to attach the initiating system and function it as he desires, giving him complete control of the operation. Also, a detonating cord loop is useful in attaching two or or more ring mains to a single priming assembly; and a single electric or nonelectric blasting cap properly fastened to two detonating cord mains by a string, wire, or piece of cloth will detonate both (fig. 86).

b. Detonating Cord Connections. A detonating cord clip (fig. 26) or square knot pulled tight is used to splice the ends of the detonating cord. At least a 6-inch length should be left free at both sides of the knot (fig. 87). When fabric is used to cover the detonating cord, the fabric must not be removed. The knot should not be placed in water or in the ground unless the charge is to be fired immediately.

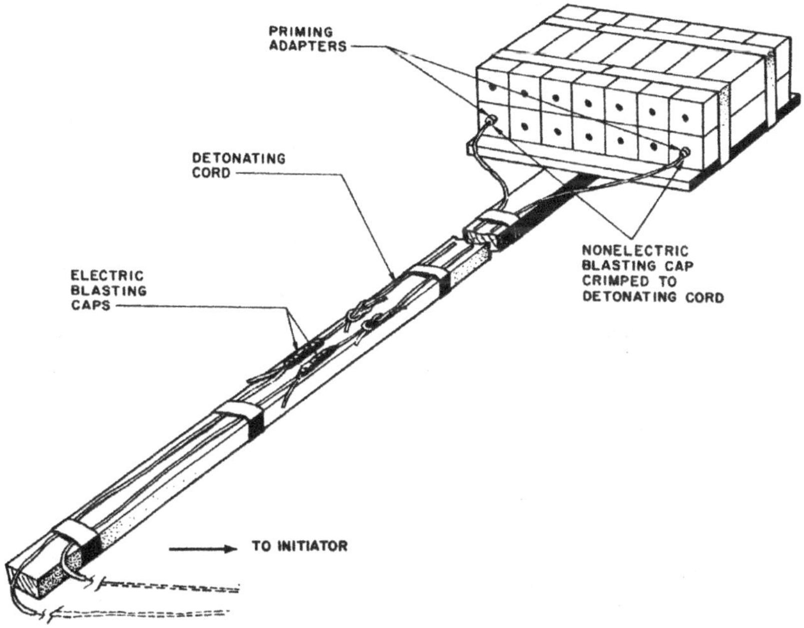

Figure 84. Detonating cord priming of pole charge.

c. Branch Line Connections. A branch line is fastened to a main line by means of a clip (fig. 26) or a girth hitch with one extra turn (fig. 88). The angle formed by the branch line and the cap end of the main line should not be less than 90° from the direction from which the blast is coming; at a smaller angle, the branch line may be blown off the main line without being detonated. At least 6 inches of the running end of the branch line is left free beyond the tie.

d. Ring Main. A ring main is made by bringing the main line back in the form of a loop and attaching it to itself with a girth hitch and one extra turn (fig. 89). This, if of sufficient length, will detonate an almost unlimited number of charges. The ring main makes the detonation of all charges more positive because the detonating wave approaches the branch lines from both directions and the charges will be detonated even when there is one break in the ring main. Branch lines coming from a ring main should be at a 90° angle. Kinks in lines should be avoided and curves and angles should be gentle. Any number of branch lines may be connected to the main line, but a branch line is never connected at a point where the main line is spliced. In making detonating cord branch line connections, avoid crossing lines. However, if this is necessary, be sure that the detonating cords are at least a foot apart in places where they cross, or they will cut each other and possibly destroy the firing system.

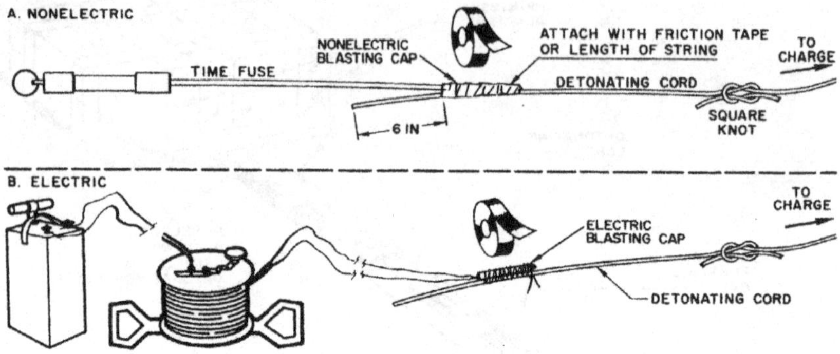

Figure 85. Electric and nonelectric assemblies attached to main line.

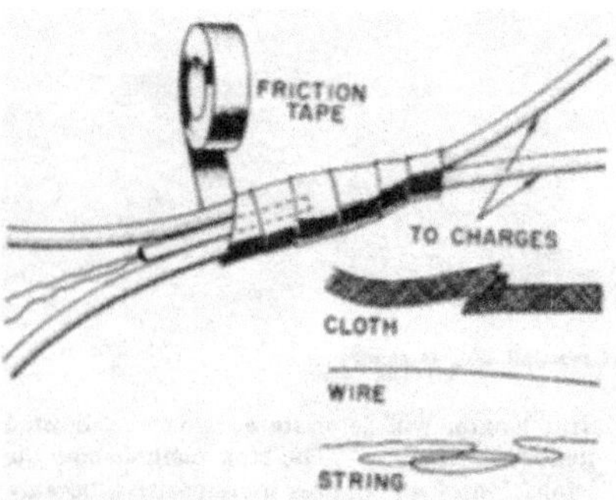

Figure 86. Fastening two lines of detonating cord to a single blasting cap assembly.

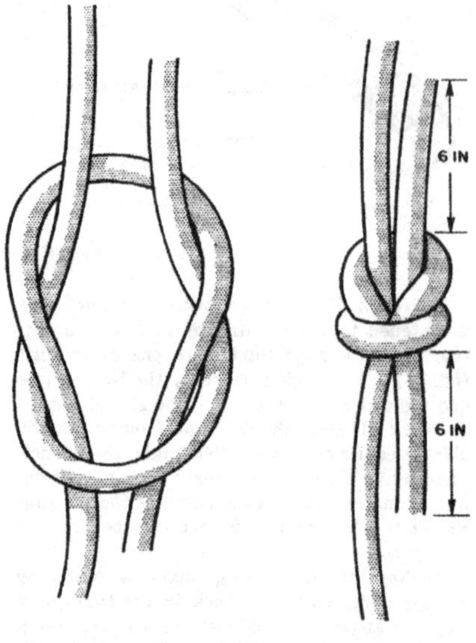

Figure 87. Square knot connections.

71. Handling Detonating Cord Misfires

a. Failure of Nonelectric Blasting Cap. If a nonelectric blasting cap initiator attached to detonating cord fails to function, delay the investigation for at least 30 minutes. Then cut the detonating cord main line between the blasting cap and the charge, and fasten a new blasting cap initiator on the detonating cord.

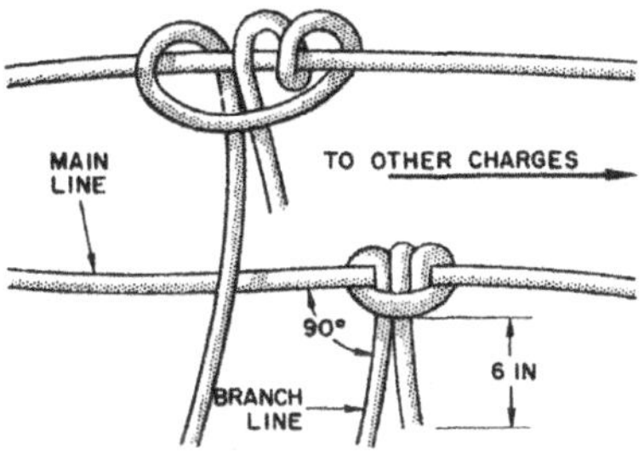

Figure 88. Girth hitch with one extra turn connecting branch line to main line.

b. Failure of Electric Blasting Cap. If an exposed electric blasting cap fastened to detonating cord fails to fire, disconnect the blasting machine immediately and investigate. Test the blasting circuit for any breaks or short circuit. If necessary, replace the original blasting cap.

c. Failure of Detonating Cord. If detonating cord fails to function at the explosion of an exposed electric or nonelectric blasting cap, investigate immediately. Attach a new blasting cap to the detonating cord, taking care to fasten it properly.

d. Failure of Branch Line. If the detonating cord main line detonates but a branch line fails, fasten a blasting cap to the branch line and fire it separately.

e. Failure of Charge to Explode. If the detonating cord leading to a charge detonates but the charge fails to explode, when above ground, delay investigation until it is certain that the charge is not burning; but when below ground, wait 30 minutes. If the charge is intact, insert a new primer. If the charge is scattered by the detonation of the original detonating cord, reassemble as much of the original charge as possible, place a new charge if necessary, and reprime. Make every attempt possible to recover all explosives scattered by misfire, particularly those used in training exercises.

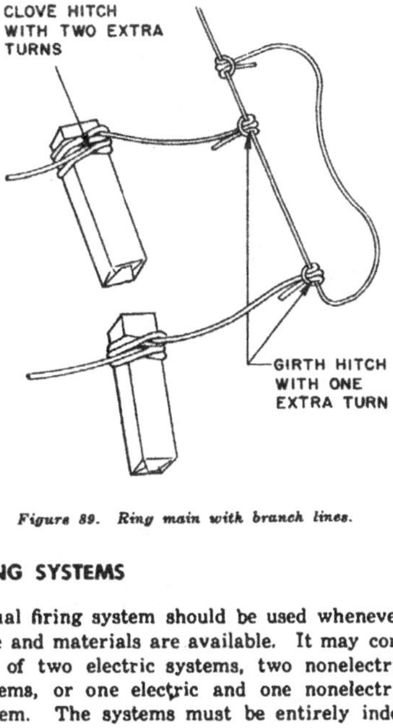

Figure 89. Ring main with branch lines.

Section IV. DUAL FIRING SYSTEMS

72. Reduction of Misfire Risks

a. The use of a dual firing system greatly increases the probability of successful firing. In combat, misfires may cause the loss of battles; in training, they cause the loss of valuable time and endanger the lives of those that investigate them. It is necessary to take every possible precaution to avoid misfires of demolition charges.

b. The failure of firing circuits is most frequently the cause of demolition misfires. Thus a dual firing system should be used whenever time and materials are available. It may consist of two electric systems, two nonelectric systems, or one electric and one nonelectric system. The systems must be entirely independent of each other and capable of firing the same charge.

73. Nonelectric Dual Firing System

This consists of two independent nonelectric systems for firing a single charge or set of

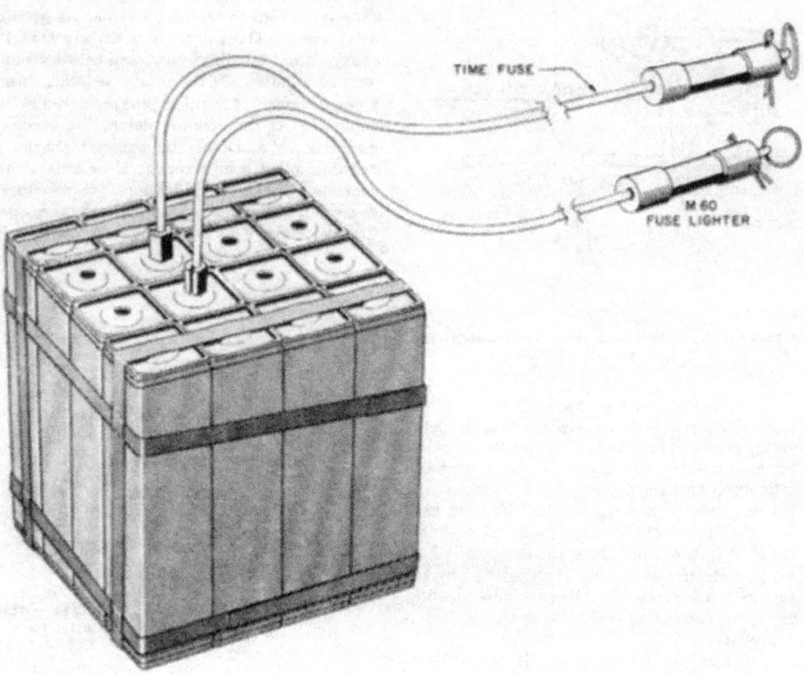

Figure 90. Nonelectric dual firing system.

charges. If two or more charges are to be fired simultaneously, two detonating cord ring mains are laid out, and a branch line from each charge is tied into each ring main. Figures 90 and 91 show the layouts for nonelectric dual firing systems.

74. Electric Dual Firing System

This dual firing system consists of two independent electric circuits, each with an electric blasting cap in each charge, so that the firing of either circuit will detonate all charges. Thus, each charge must have two electric primers. The correct layout is shown in figure 92. The firing wires of the two circuits should be kept separated so that both will not be cut by a single bullet or a single shell fragment. The firing points also should be at two separate locations.

75. Combination Dual Firing System

The combination dual firing system requires an electric and nonelectric firing circuit (fig. 93). Each charge has an electric and nonelectric primer. Both the electric and nonelectric circuits must be entirely independent of each other. The nonelectric system should be fired first.

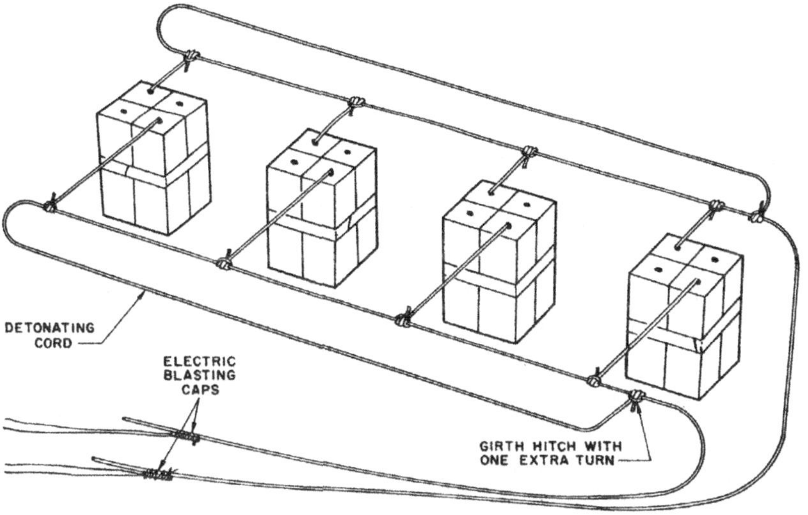

Figure 91. Dual detonating cord priming, using two ring mains.

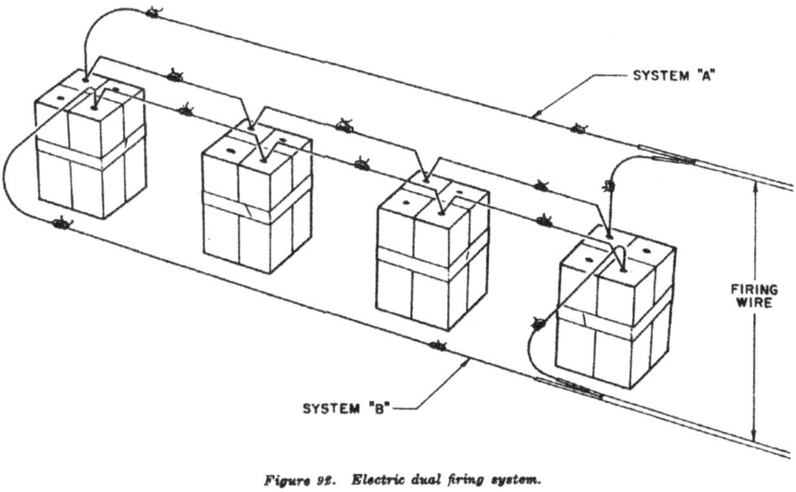

Figure 92. Electric dual firing system.

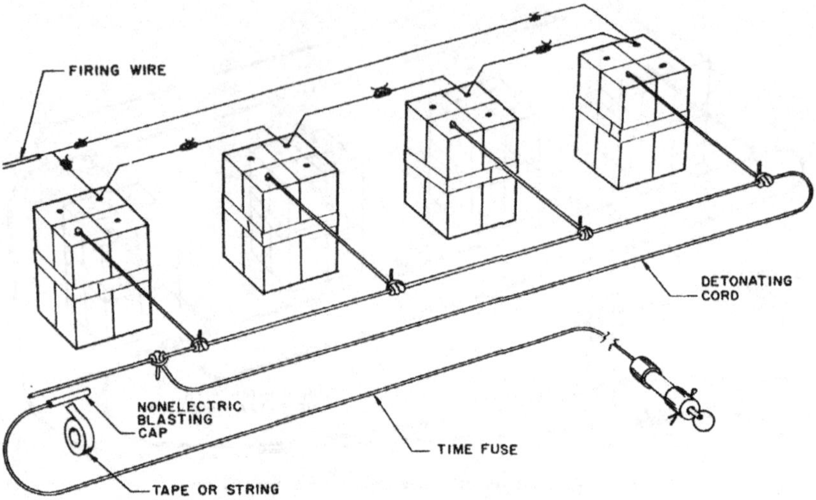

Figure 93. Combination dual firing system.

CHAPTER 3

CALCULATION AND PLACEMENT OF CHARGES

Section I. INTRODUCTION

76. Critical Factors in Demolitions

The critical factors in demolitions are the type of explosive used and the size, placement, and tamping of the charge.

a. Type of Explosive. Explosives used in military operations and their application to demolition projects are shown in table VIII.

Table VIII. Characteristics of Explosives

Name	Principal use	Smallest cap * required for detonation	Approx. velocity of detonation (meter/sec) (feet/sec)	Relative effectiveness as external charge (TNT-1.00)	Intensity of poisonous fumes	Water resistance
TNT	Main charge, booster charge, cutting and breaching charge, general and military use in forward areas	Special blasting cap	6,900 mps 23,000 fps	1.00	Dangerous	Excellent
Tetrytol			7,000 mps 23,000 fps	1.20	Dangerous	Excellent
Composition C 3			7,625 mps 25,018 fps	1.34	Dangerous	Good
Composition C 4			8,040 mps 26,379 fps	1.34	Slight	Excellent
Ammonium Nitrate	Cratering and ditching		3,400 mps 11,000 fps	0.42	Dangerous	Poor
Military Dynamite M1	Quarry and rock cuts		6,100 mps 20,000 fps	0.92	Dangerous	Good
Straight Dynamite (commercial) 40%	Land clearing, cratering quarrying, and general use in rear areas	No. 6 commercial cap	4,600 mps 15,000 fps	0.65	Dangerous	Good (if fired within 24 hours)
50%			5,500 mps 18,000 fps	0.79		
60%			5,800 mps 19,000 fps	0.83		
Ammonia Dynamite (commercial) 40%	Land clearing, cratering quarrying, and general use in rear areas	No. 6 commercial cap	2,700 mps 8,900 fps	0.41	Dangerous	Poor
50%			3,400 mps 11,000 fps	0.46		
60%			3,700 mps 12,000 fps	0.53		
Gelatin Dynamite (commercial) 40%	Land clearing, cratering quarrying, and general use in rear areas	No. 6 commercial cap	2,400 mps 7,900 fps	0.42	Slight	Good
50%			2,700 mps 8,900 fps	0.47		
60%			4,900 mps 16,000 fps	0.76		

Table VIII. Characteristics of Explosives—Continued

Name	Principal use	Smallest cap * required for detonation	Approx. velocity of detonation (meter/sec) (feet/sec)	Relative effectiveness as external charge (TNT-1.00)	Intensity of poisonous fumes	Water resistance
Ammonia Gelatin Dynamite (commercial) 40% 60%	Land clearing, cratering quarrying, and general use in rear areas	No. 6 commercial cap	4,900 mps 16,000 fps 5,700 mps 18,700 fps		Slight	Excellent
PETN	Detonating cord Blasting caps	Special blasting cap N/A	7,300 mps 24,000 fps	1.66	Slight	Good
TETRYL	Booster charge	Special blasting cap	7,100 mps 23,300 fps	1.25	Dangerous	Excellent
RDX	Base charge in blasting caps		8,350 mps 27,440 fps	1.50		
Composition B	Shaped charges	Special blasting cap	7,800 mps 25,600 fps	1.35	Dangerous	Excellent
Amatol 80/20	Bangalore torpedo	Special blasting cap	4,900 mps 16,000 fps	1.17	Dangerous	Poor
Black Powder	Time blasting fuze	N/A	400 mps 1,312 fps	0.55	Dangerous	Poor

* Electric or nonelectric

Abbreviations: fps — feet per second; mps — meters per second; N/A — not applicable; No. — number; sec. — second(s)

b. Size of Charge. The amount of explosive used in a demolition project is determined by formula calculation, and by means of a computing tape or tables. Formulas for computing specific charges—timber and steel cutting, breaching, and so on—are given in succeeding sections of this chapter. In the formulas (for example, $P = D^2$), the value of P is the amount of TNT (in pounds) required for external charges. If other explosives are used, the value of P must be substituted according to the strength of these other explosives in relation to TNT. The substitution is computed by dividing the P value (TNT) by the relative effectiveness factor for the explosive to be used. Steel and timber charges should be computed by formula when possible.

c. Tamping. The detonation of an explosive produces pressure in all directions. If the charge is not completely sealed in or confined or of the material surrounding the explosive is not equally strong on all sides, the explosive force breaks through the weakest spot and part of the destructive effect is lost. To retain as much of this explosive force as possible, material is packed around the charge. This material is called *tamping material or tamping,* and the process, *tamping.* On the other hand, an internal charge (one placed in the target to be destroyed) is confined by packing material in the borehole on top of the charge as is done in quarrying and cratering. This is called *stemming.* Explosive charges are generally tamped and stemmed as described below.

d. Charge Placement.

(1) Charges should be placed at the position that will provide maximum effectiveness. For cratering, they are placed in holes in the ground; for breaking or collapsing stone or concrete, they are properly located on the surface or in boreholes; for cutting standing timber they may be tied on the outside or placed in boreholes, whichever is the more practical.

(2) Charges are fastened to the target by

wire, adhesive compound, friction tape, or string; propped against the target by means of a wooden or metal frame made of scrap or other available materials; or placed in boreholes. Special accessories are issued for this purpose—adhesive compound, the rivet-punching powder-actuated driver, the earth auger, and pneumatic tools (para 29–41).

77. Types of Charges

a. Internal Charges. Internal charges are charges placed in boreholes in the target. These are confined by tightly packing sand, wet clay, or other material (stemming) into the opening. This is tamped and packed against the explosive to fill the hole all the way to the surface. In drill holes, tamping should not begin until the explosive is covered by at least one foot of stemming. Light materials are not acceptable, as they are apt to blow out of the borehole and cause incomplete destruction; neither are flammable materials like paper, sawdust, and sacking, which may ignite.

b. External Charges. These charges are placed on the surface of the target. They are tamped by covering them with tightly packed sand, clay, or other dense material. Tamping may be in sandbags or loose. For maximum effectiveness the thickness of the tamping should at least equal the breaching radius. Small breaching charges on horizontal surfaces are sometimes tamped by packing several inches of wet clay or mud around them. This process is called mudcapping.

Section II. TIMBER-CUTTING CHARGES

78. Size and Placement of Charge

For untamped external charges, block explosive (TNT, tetrytol, and Composition C3 and C4) is adequate, as it is easily tied or fastened in place and the charge size is calculated by formula based on its effectiveness in relation to that of TNT (relative effectiveness factor). For tamped internal charges in boreholes, dynamite is generally used, as it is the most convenient to place because of the size of the cartridge and is powerful enough because it is confined. It is impractical to attempt to cut all kinds of timber with charges of a size calculated from a single formula. There is too much variation in different kinds of timber from locality to locality. Accordingly, test shots must be made to determine the size of the charge to cut a specific type of timber. Formulas for the calculation of these shots are provided for untamped external charges, felling trees for an abatis, and for tamped internal charges. They are as follows:

a. Formula for Untamped External Charges. For cutting trees, piles, posts, beams, or other timber members with an untamped external charge, the following formula is used for the test shot for either round or rectangular members:

$$P = \frac{D^2}{40}$$

P = pounds of TNT required,

P = least diameter of the timber in inches or the least dimension of dressed timber, and

40 = constant

Adjustment for explosives other than TNT will be made using the relative effectiveness factor (table VIII) that pertains to the particular explosive being used. The amount of explosive required to cut a round timber 30 inches in diameter using an untamped external charge is determined as follows:

$$P = \frac{D^2}{40}$$

D = 30 inches

$$P = \frac{(30)^2}{40} = 22.5 \text{ pounds of TNT.}$$

b. Formula for Partially Cutting Trees to Create an Obstacle or Abatis. When cutting trees and leaving them attached to the stumps to create an obstacle, the formula $P = \frac{D^2}{50}$ is used to compute the amount of TNT required for the test shot. The result of the test shot will determine the need for increasing or decreasing the amount of explosive required for subsequent shots

c. Formula For Tamped Internal Charges. Tamped internal cutting charges may be calculated by the following formula:

$$P = \frac{D^2}{250}$$

P = pounds of explosive required,
D = diameter or least dimension of dressed timber, in inches, and
250 = constant

Note. P = any explosive; relative effectiveness factor is not pertinent.

The amount of explosive required to cut a 15-inch diameter tree, using tamped internal charges, is determined as follows:

$$P = \frac{D^2}{250}$$

$$P = \frac{15^2}{250} = \frac{225}{250} = 0.90 \text{ or } 1 \text{ pound}$$

Note. See rounding off rule para 81*b*(4).

d. External Charge Placement. External charges are placed as close as possible to the surface of the timber regardless of the kind of cut desired (fig. 94). Frequently it is desirable to notch the tree to hold the explosive in place. If the tree or timber is not round and the direction of fall is of no concern, the explosive is placed on the widest face so that the cut will be through the least thickness. The tree will fall toward the side where the explosive is placed, unless influenced by lean or wind. To make the direction of fall certain a "kicker charge," a one-pound block of TNT, placed about two-thirds of the distance up the tree on the opposite side has proved excellent (fig. 94). Charges on rectangular or square dressed timber are placed as shown in figure 95.

e. Internal Charge Placement. These charges are placed in boreholds parallel to the greatest dimension of cross section and tightly tamped with moist clay or sand. If the charge is too large to be placed in one borehole, bore two side by side. On round timber, bore the two holes at approximately right angles to each other, but not to intersect (fig. 96). Both boreholes are tamped and the charges are fired simultaneously.

Figure 94. External timber-cutting charge.

79. Abatis

Charges for making fallen-tree obstacles are placed the same as those in paragraph 78, except that they are placed approximately 5 feet above ground level. To make the obstacles more difficult to remove, they should be mined, boobytrapped, and covered by fire. To be effective these obstacles should be at least 75 meters in depth and the felled trees should extend at a 45° angle toward the enemy.

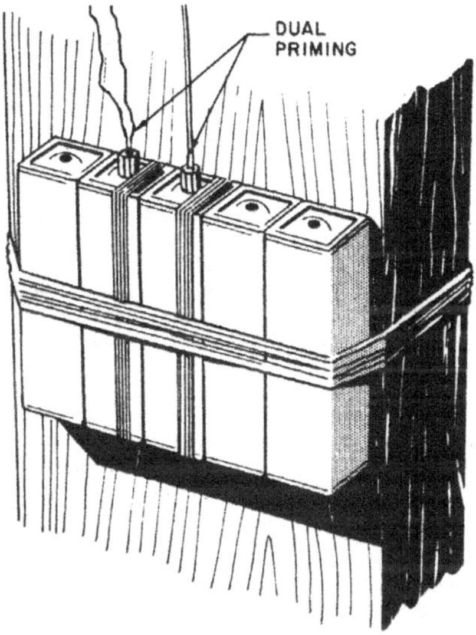

Figure 95. External cutting charge on rectangular timber.

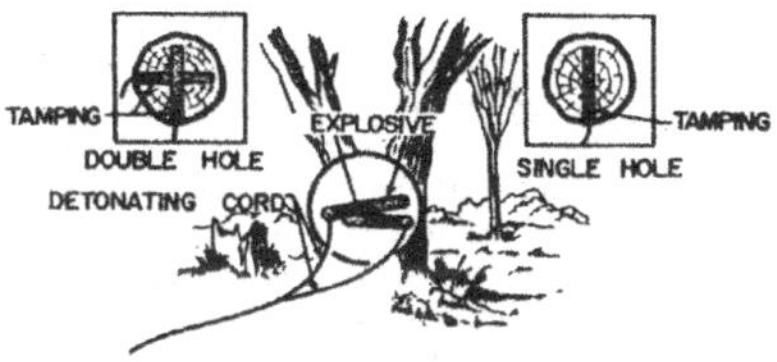

Figure 96. Internal timber-cutting charge.

Section III. STEEL-CUTTING CHARGES

80. Cutting Steel with Explosives

a. Principles of Detonation.

(1) When a high explosive detonates, the explosive changes violently from a solid into compressed gas at extremely high pressure. The rate of change is determined among other things by the type of explosive and the density, confinement, and the dimensions of the charge. Thus the detonation releases tremendous pressure in the form of a shock wave which, although it exists for only a few micro-seconds at any given point, may shatter and displace objects in its path as it proceeds from its point of origin. This shock wave is transmitted directly to any substance in contact with the charge, other characteristics being equal.

(2) A high explosive charge detonated in direct contact with a steel plate produces easily detectable destructive effects. An indentation or depression with an area about the size of the contact area of the explosive charge is made in the surface of the plate where the charge is exploded. A slab of metal—a spall or scab—is torn from the free surface of the plate directly opposite the explosive charge. This spalled metal is approximately the shape of the explosive charge, but

its area is usually greater than the contact area of the charge. The steel is split or fractured under the exploded charge along its entire length, and finally, a cross fracture is formed across the end of the charge away from the point of initiation. Variations in the dimensions of the charge, the shape of the steel member under attack, and the placement of the charge in relation to the steel member can alter the destructive effects described.

b. Significance of Charge Dimensions.

(1) The force of an explosion is proportional to the quantity and power of the explosive, but the destructive effect depends on the contact between the explosive and the target and on the manner that the explosive is directed at the target. For the maximum destructive effect against a steel target, an explosive charge with a configuration and dimensions optimum for the size and shape of the target must be detonated in intimate contact with the steel along the desired line of cut. As any air or water gap between the charge and the steel greatly reduces the cutting effect, clóse contact is essential. An optimum relation must exist between the area of the charge in contact with the target and charge thickness in order to transmit the greatest shock. If any given weight of explosive, calculated to cut a given target, is spread too thinly, there will be insufficient space for the detonation to attain full velocity before striking the target. The shock wave will tend to travel more nearly parallel than normal to the surface over much of the area; and the volume of the target will be excessive for the strength of the shock wave. On the other extreme, a thick charge with narrow contact area will transmit the shock wave over too little of the target with excessive lateral loss of energy. Test results have demonstrated conclusively that the optimum ratio of charge width to charge thickness is about 3:1 for contact steel cutting charges placed on structural members 3 inches thick or less. They have also demonstrated that the point of charge initiation has no significant effect on the shattering power of contact charges on steel. Successful explosive cutting of steel bars and structural steel beams with certain contact charges, however, requires charge initiation at specific points. These findings are applied in the discussion on steel-cutting formulas given below.

(2) Thus, in the preparation of steel cutting charges, the factors of type, size, and placement of the explosive are important for successful operations. The confinement or tamping of the charge is rarely practical or possible. Formulas for the computation of the size of the charge vary with the types of steel—structural, high carbon, and so forth. Placement, frequently harder to accomplish on steel structures than on others, is aided by the use of plastic explosive and the M118 special sheet explosive.

c. Explosive Used. Steel cutting charges are selected because of their cutting effect and adaptability to placement. Plastic explosive (C4) and sheet explosive M118 are the most desirable as they have high detonation velocity and other characteristics that give them great cutting power. C4 (M5A1) can be molded or cut to fit tightly into the grooves and angles of the target, particularly structural steel, chains, and steel cables; but the M112 block, also C4, is more adaptable because of its adhesive compound on one face, which fixes it more securely on the target. Sheet explosive, because of its width (3 in.), thickness (1/4 in.), and adhesive, is more desirable for some steel targets than either the M5A1 or M112 demolition block. TNT, on the other hand, is adequate, generally available, and cast into blocks that may be readily assembled and fixed, but not molded to the target.

81. Size of Charge Determined by Type and Size of Steel

a. Types of Steel.

(1) *Structural.* Examples of this are I-beams, wide-flanged beams, channels, angle sections, structural tees, and steel plates used in building or bridge construction. These are the types of steel usually present in demolition projects. The formula in *b*(1), below, is applicable to structural steel, except for cutting slender structural bars (2 inches or less in diameter) where placement difficulties require the use of the formula in *b*(2)(*b*) below.

(2) *High-carbon.* This type of steel is used in the construction of metal-working dies and rolls. The formula in *b*(2)(*a*), below, is applicable.

(3) *Alloy.* Gears, shafts, tools, and plowshares generally are made of alloy steel. Chains and cables also are often made from alloy steel; some, however, are made of a high-carbon steel. The formula in *b*(2)(*a*), below, applies to high carbon or alloy steel.

(4) *Nickel-molybdenum steel and cast iron.* Cast iron, being very brittle, breaks easily, but nickel-molybdenum steel cannot be cut by demolition blocks. The jet from a shaped charge will penetrate it, but cutting will require multiple charges. Accordingly, nickel-molybdenum steel should be cut by some method other than explosives—acetylene or electric cutting tools, for example.

b. Calculation of Charges.

(1) *Formula for structural steel.* Charges to cut I-beams, builtup girders, steel plates, columns, and other structural steel sections are computed by formula as follows:

$$P = \frac{3}{8} A$$

P = pounds of TNT required,
A = cross-section area, in square inches, of the steel member to be cut, and

$$\frac{3}{8} = \text{constant}$$

(2) *Formula for other steels.*

(*a*) The formula below is recommended for the computation of cutting charges for high-carbon or alloy steel, such as that found in machinery.

$P = D^2$
P = pounds of TNT
D = diameter or thickness in inches of section to be cut.

(*b*) For round steel bars, such as concrete reinforcing rods, where the small size makes charge placement difficult or impossible and for chains, cables, strong forgings, steel rods, machine parts, and high-strength tools of a diameter of 2 inches or less use—

P = D
P = pounds of TNT
D = diameter in inches of section to be cut.

Such steel, however, may be cut by "rule of thumb:"

For round bars up to 1 inch in diameter, use 1 pound TNT.

For round bars over 1 inch up to 2 inches in diameter, use 2 pounds of TNT.

(3) *Railroad rail.* The size of railroad rail is usually expressed in terms of weight per yard. Rails over 80 pounds per yard (more than 5 inches in height) may be cut with 1 pound of TNT. For rails less than 80 pounds per yard (5 inches or less in height), ½ pound of TNT is adequate.

(4) *"Rounding-off" rule.* Charges calculated by formulas should be "rounded-off" to the next higher unit package of explosive. However, when a ½-pound charge is required, and only 2½-pound blocks are available, cut the blocks to proper size if feasible. For charges other than TNT, apply the

"rounding off" rule at completion of relative effectiveness factor calculations.

(5) *Problem: cutting steel I-beam.* Determine the amount of TNT required to cut the steel I-beam shown in figure 97. The solution is given in the figure.

(6) *Problem: explosives other than TNT.* Determine the amount of C4 explosive required to cut the steel I-beam in figure 97.

$$\text{The amount of TNT} = 3\ \frac{27}{64}$$

As C4 is 1.34 times as effective as TNT

$$\text{P (of C4)} = \frac{\frac{219}{64}}{1.34} = 2.5 \text{ pounds}$$

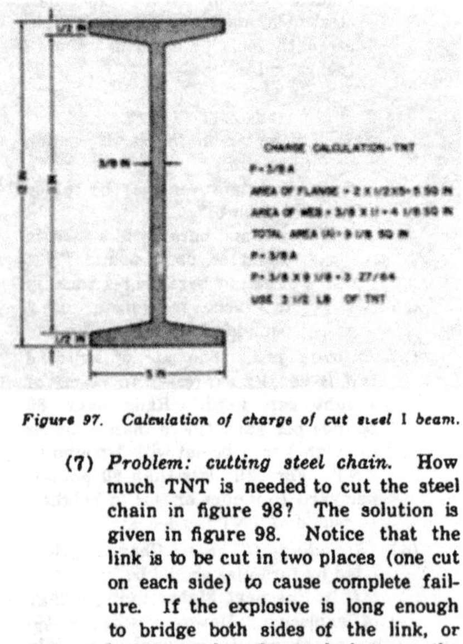

Figure 97. Calculation of charge of cut steel I beam.

(7) *Problem: cutting steel chain.* How much TNT is needed to cut the steel chain in figure 98? The solution is given in figure 98. Notice that the link is to be cut in two places (one cut on each side) to cause complete failure. If the explosive is long enough to bridge both sides of the link, or large enough to fit snugly between the two sides, use only one charge; but if it is not, use two separately-primed charges.

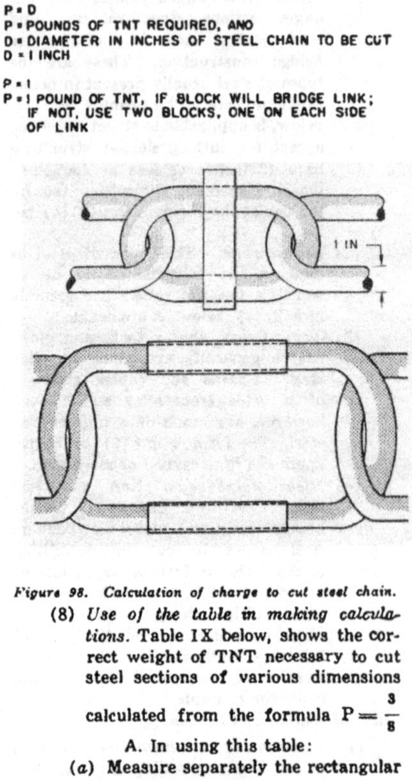

Figure 98. Calculation of charge to cut steel chain.

(8) *Use of the table in making calculations.* Table IX below, shows the correct weight of TNT necessary to cut steel sections of various dimensions calculated from the formula $P = \frac{3}{8}$

A. In using this table:

(*a*) Measure separately the rectangular sections of members.

(*b*) Find the corresponding charge for each section by using the table.

(*c*) Total the charges for the sections.

(*d*) Use the next larger given dimension if dimensions of sections do not appear in the table.

Caution: Never use less than the calculated amount.

Table IX. TNT Needed to Cut Steel Sections

Average thickness of section in inches	Pounds of explosive * for rectangular steel sections of given dimensions																
	Height of section in inches																
	2	3	4	5	6	7	8	9	10	11	12	14	16	18	20	22	24
¼	0.2	0.3	0.4	0.5	0.6	0.7	0.8	0.9	1.0	1.1	1.2	1.3	1.5	1.7	1.9	2.1	2.3
⅜	0.3	0.5	0.6	0.7	0.9	1.1	1.2	1.3	1.4	1.6	1.7	2.0	2.3	2.6	2.8	3.1	3.4
½	0.4	0.6	0.8	1.0	1.2	1.4	1.5	1.7	1.9	2.1	2.3	2.7	3.0	3.4	3.8	4.2	4.5
⅝	0.5	0.7	1.0	1.2	1.4	1.7	1.9	2.2	2.4	2.7	2.9	3.3	3.8	4.3	4.7	5.2	5.7
¾	0.6	0.9	1.2	1.4	1.7	2.0	2.3	2.6	2.8	3.1	3.4	4.0	4.5	5.1	5.7	6.3	6.8
⅞	0.7	1.0	1.4	1.7	2.0	2.4	2.7	3.0	3.3	3.7	4.0	4.6	5.3	6.0	6.6	7.3	7.9
1	0.8	1.2	1.5	1.9	2.3	2.7	3.0	3.4	3.8	4.2	4.5	5.3	6.0	6.8	7.5	8.3	9.0

* TNT

To use:
1. Measure rectangular sections of members separately.
2. Using table, find charge for each section.
3. Add charges for sections to find total charge.
4. If dimensions of sections do not appear on table, use the next larger dimension.

e. Problem. The problem in figure 97 may be solved as follows:

Charge for flanges:
width = 5 inches
thickness = ½ inch
Charge from table =
1.0 pounds

Charge for web:
height = 11 inches
thickness = ⅜ inch
Charge from table =
1.6 pounds

Total charge: 2 flanges = 2 x 1.0 = 2.0 pounds
web = 1 x 1.6 = 1.6 pounds
Use 4 pounds of TNT. 3.6 pounds

82. Charge Placement

a. Steel Sections. The size and type of a steel section determine the placement of the explosive charge. Some elongated sections may be cut by placing the explosive on one side of the section completely along the proposed line of rupture. In some steel trusses in which the individual members are fabricated from two or more primary sections, such as angle irons or bars separated by spacer washers or gusset plates, the charge has to be cut with the opposing portions of the charge slightly offset to produce a shearing action (para 83*d*(4). Heavier H-beams, wide flange beams, and columns may also require auxiliary charges placed on the outside of the flanges. Care must be taken to insure that opposing charges are never directly opposite each other, otherwise they tend to neutralize the explosive effect.

b. Rods, Chains, and Cables. Block explosive, often difficult to emplace, is not recommended for cutting steel rods, chains, and cables if plastic explosive is available.

c. Steel Members and Railroad Rails. Charge placement for cutting these is found in figures 99 and 140.

d. Built-up Members. Built-up members frequently have an irregular shape, which makes it difficult to obtain a close contact between the explosive charge and all of the surface. If it is impractical to distribute the charge properly to obtain close contact, the amount of explosive should be increased.

e. Irregular Steel Shapes. Composition C4 (M2A1 block) is a good explosive for cutting irregular steel shapes because it is easily molded or pressed into place to give maximum contact. A light coating of adhesive compound applied to the steel surface will help hold the explosive on the target. The M112 block, also C4, has an adhesive coating on one side and the M118 sheet explosive has a similar coating, which makes placement easier (para 80c).

f. Securing Explosives in Place. All ex-

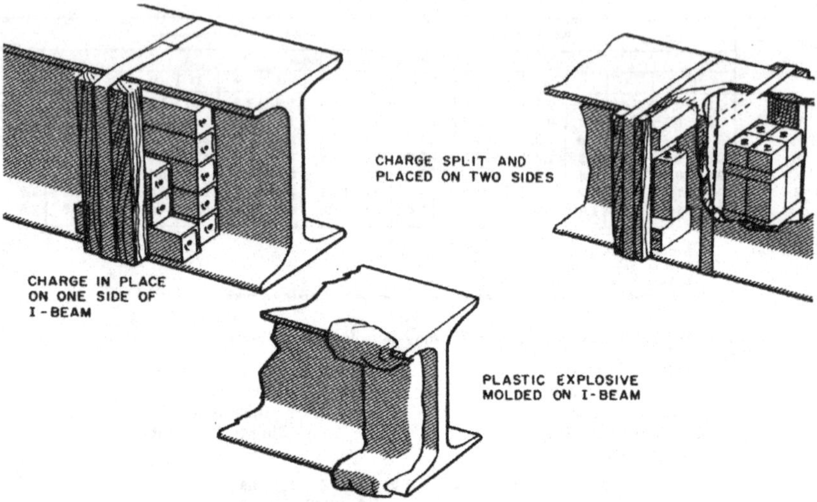

Figure 99. Placement of charges on steel members.

plosives except moldable or adhesive types must be tied, taped, or wedged in place unless they rest on horizontal surfaces and are not in danger of being jarred out of place.

g. Precautions. In cutting steel, the charge should be placed on the same side as the firing party, as explosive charges throw steel fragments (missiles) long distance at high velocities.

83. Special Steel Cutting Techniques

a. Use of Special Techniques. Three types of steel cutting charges are available for use, the saddle charge, diamond charge, and ribbon charge. They are prepared in advance for transportation to the site by wrapping them in aluminum foil or heavy paper. Not more than one thickness of the wrapper should be between the explosive and the target.

b. Saddle Charge. This charge is used on solid cylindrical mild steel targets up to 8 inches in diameter. Detonation is initiated at the apex of the long axis (fig. 100).

(1) *Size of charge.*

(*a*) Thickness of charge = 1/3 of the thickness of M5A1 block of plastic explosive (2/3 inch) for targets up to 19 inches in circumference (6 inches in diameter); and ½ the thickness of M5A1 block of plastic explosive (1 inch) for targets from 19 to 25 inches in circumference (over 6 to 8 inches in diameter).

(*b*) Base of charge = ½ circumference of target.

(*c*) Long axis of charge = circumference of target.

(2) *Example.* Determine the dimensions of a charge for cutting a shaft 18 inches in circumference (may be measured by a length of string).

(*a*) Thickness = 1/3 thickness of M5A1 block of plastic explosive (2/3 inch).

(*b*) Base = ½ x 18 = 9 inches

(*c*) Long axis = 18 inches.

Charge is 9 inches at base, 18 inches at long axis, and 2/3 inch thick.

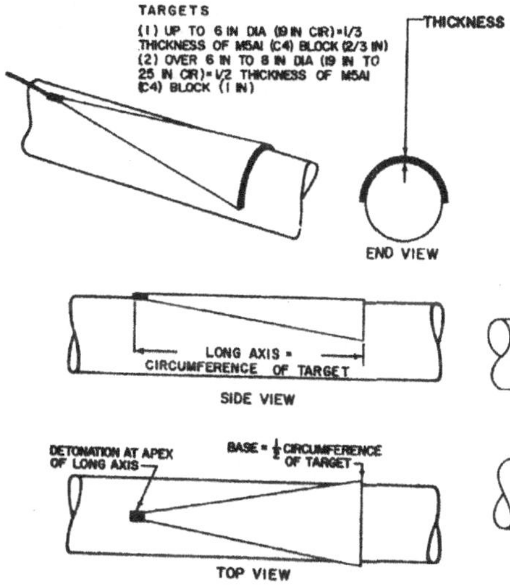

Figure 100. Saddle charge.

Note. Steel alloy and mild steel targets over 25 inches in circumference (over 8 inches in diameter) require the diamond charge (c below).

(3) *Placement.* The long axis of the saddle should be parallel with the long axis of the target. Detonation of the charge is by the placement of a military electric or nonelectric blasting cap at the apex of the long axis.

c. Diamond Charge. This is used on high carbon steel or steel alloy targets (fig. 101). It is shaped like a diamond.

(1) *Size of charge.* The size of the charge depends on the dimensions of the target.

(*a*) Long axis of charge = circumference of the target.

(*b*) Short axis of charge = ½ the circumference of the target.

(*c*) Thickness of charge = 1/3 the thickness of M5A1 block of plastic explosive (2/3 inch).

(2) *Example.* Determine the size of a charge for cutting a steel alloy shaft 15 inches in circumference.

(*a*) Long axis = 15 inches.

(*b*) Short axis = ½ x 15 = 7½ inches.

(*c*) Thickness = 1/3 the thickness of M5A1 block of plastic explosive or 2/3 inch. Charge is 15 inches at long axis, 7½ at short axis, and 2/3 inch thick.

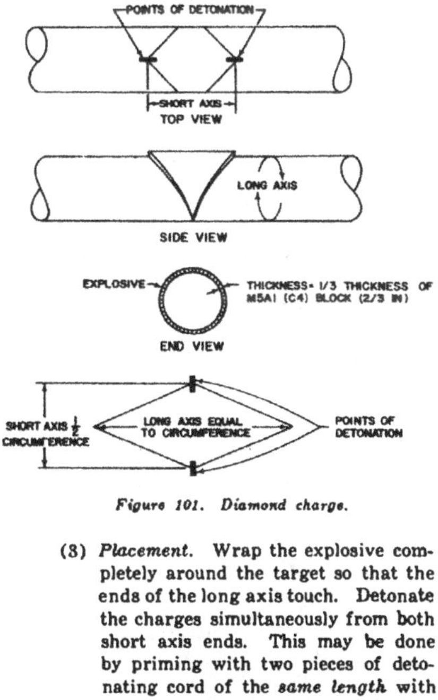

Figure 101. Diamond charge.

(3) *Placement.* Wrap the explosive completely around the target so that the ends of the long axis touch. Detonate the charges simultaneously from both short axis ends. This may be done by priming with two pieces of detonating cord of the *same length* with nonelectric blasting caps crimped to the ends, or two electric blasting caps connected in series.

d. Ribbon Charge. This charge, if properly calculated and placed, cuts steel with considerably less explosive than standard charges. It is effective on noncircular steel targets up to 2 inches thick (fig. 102).

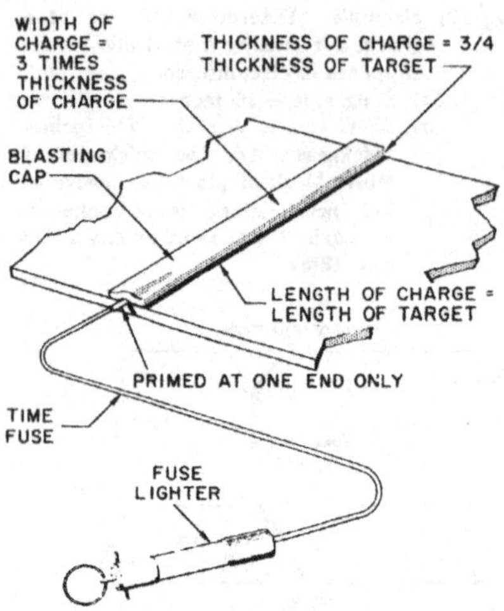

Figure 102. Ribbon charge.

(1) *Calculation.* The effectiveness of the blast depends on the width and thickness of the explosive.

(*a*) Thickness of charge = ¾ thickness of the target.

(*b*) Width of charge = 3 x the thickness of the charge.

(*c*) Length of charge = length of the cut.

Note. Charge should never be less than ⅛-inch thick.

(2) *Example.* Determine the thickness and width of a ribbon charge for cutting a steel plate 1 inch thick.

(*a*) Thickness = ¾ thickness of target
1 x ¾ = ¾ inch.

(*b*) Width = 3 x thickness of the charge
3 x ¾ = 2¼ inches. Charge is ¾ inches thick and 2¼ inches wide.

(3) *Detonation.* The ribbon charge is detonated from one end only. It may be necessary where the calculated thickness is small to "build up" the detonating end with extra explosive. Either the electric or nonelectric cap is satisfactory. The charge should extend slightly over the target at each end to insure a complete cut. The steel member will be ruptured at approximately the linear axis of the ribbon.

(4) *Use on structural steel sections.* The ribbon charge (computed by formula given in (2), above) has a proven application to cutting structural steel sections (fig. 103). On wide-flanged beams of less than 2 inches of steel thickness, a C-shaped charge is placed on one side to cut the web and half of the top and bottom flanges. The other sides of these flanges are cut by two offset ribbon charges, placed so that one edge is opposite the center of the C-shaped charge as shown in A, figure 103. For beams with steel

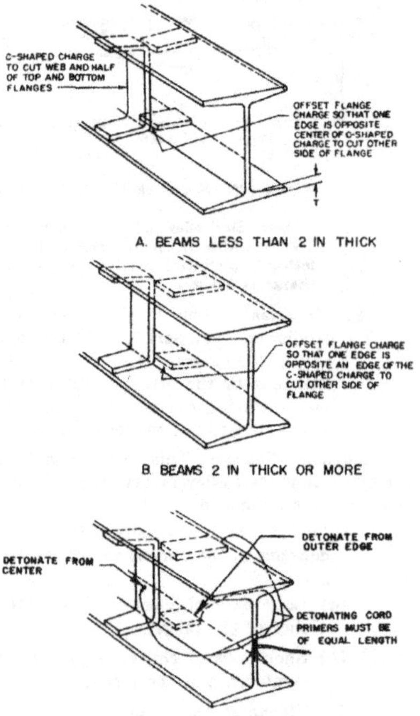

Figure 103. Placement of ribbon charges on structural steel sections.

thickness of 2 inches and over, the offset charges are placed opposite an edge of the C-shaped charge as shown in B, figure 103. For optimum results, the charges must be primed with three equal lengths of detonating cord with blasting caps attached and placed in the charges as in C, figure 103 to provide simultaneous detonation. The three charges are initiated by an electric or nonelectric system.

Note. It is possible that on heavily-loaded beams the C-shaped ribbon charge placed on one side (fig. 103) will be adequate. The uncut sides of the flanges may fail without the use of offset charges because of the weight. This technique, however, must be used with discretion to eliminate possibility of failure.

Section IV. PRESSURE CHARGES

84. Size of Charge

The pressure charge is used primarily for the demolition of simple span reinforced concrete T-beam bridges and cantilever bridges.

a. Formula for Tamped Pressure Charges. The amount of TNT required for a tamped pressure charge is calculated by the formula below. If explosive other than TNT is used, the calculated value must be adjusted.

$P = 3H^2T$

P = pounds of TNT required for each beam (stringer).

H = height of beam (including thickness of roadway) in feet

T = thickness of beam in feet

However, the values of H and T, if not whole numbers should have the fraction expressed in ¼-foot increments by rounding off to the next higher ¼-foot dimension. H and T are never considered less than 1 foot.

b. Formula for Untamped Pressure Charges. The value calculated for P by the above formula is increased by one-third if the pressure charge is not tamped to a minimum of 10 inches.

c. Problem: Pressure Charges. Determine the amount of TNT required to destroy the bridge span shown in figure 104. The solution to this problem is found in the figure. Notice that the quantity of explosive given by the formula refers to the charge for each beam. Thus four of these 41-pound charges should be placed as shown in the figure.

d. Use of Table in Making Calculations. Table X, gives the various weights of TNT required to provide suitable tamped pressure charges. The weights of TNT in the table were calculated from the formula $P = 3H^2T$ and the values were rounded off to the next highest pound. To use the table proceed as follows:

(1) Select the appropriate value in the "Height of beam" column.

(2) Read the weight of the TNT from the column corresponding to the thickness of the beam.

e. Example. The height of the beam in the problem in figure 104 is 34 inches and the thickness is 17 inches. In table X (36-inch height and 18-inch thickness columns), the weight of TNT for the tamped pressure charge is indicated as 41 pounds. For untamped charges the weight values given in the table are increased by one-third.

85. Charge Placement and Tamping

a. Placement. The correct amount of explosive is placed on the roadway over the centerline of each stringer (fig. 104) and alined midway between the ends of the span. If a curb or side rail prevents placing the charge directly above the outside stringer, it is placed against the curb or side rail. This does not require an increase in the size of the explosive charge.

b. Tamping. Pressure charges should be tamped whenever possible. Effective tamping requires a minimum of 10 inches of material. All charges are primed to fire simultaneously.

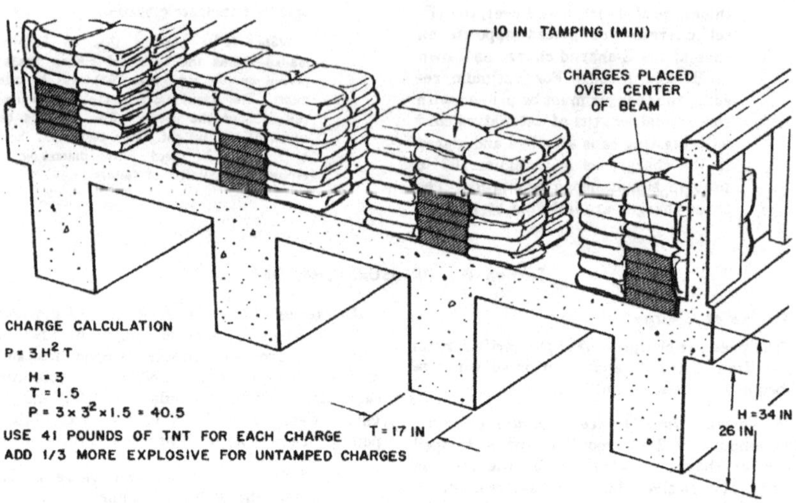

Figure 104. Calculation and placement of pressure charges.

Table X. TNT Required for Tamped Pressure Charges

Pounds of explosive for each beam (tamped charges)* (TNT)

Height of beam in feet	Thickness of beam in feet								
	1 (12 in.)	1¼ (15 in.)	1½ (18 in.)	1¾ (21 in.)	2 (24 in.)	2¼ (27 in.)	2½ (30 in.)	2¾ (33 in.)	3 (36 in.)
1 (12 in.)	3	-----	-----	-----	-----	-----	-----	-----	-----
1 ¼ (15 in.)	5	6	-----	-----	-----	-----	-----	-----	-----
1 ½ (18 in.)	7	9	11	-----	-----	-----	-----	-----	-----
1 ¾ (21 in.)	10	12	14	16	-----	-----	-----	-----	-----
2 (24 in.)	12	15	18	21	24	-----	-----	-----	-----
2 ¼ (27 in.)	16	19	23	27	31	35	-----	-----	-----
2 ½ (30 in.)	19	24	29	33	38	43	47	-----	-----
2 ¾ (33 in.)	23	29	34	40	46	51	57	63	-----
3 (36 in.)	27	34	41	48	54	61	68	75	81
3 ¼ (39 in.)	32	40	48	56	64	72	80	88	95
3 ½ (42 in.)	37	46	56	65	73	83	92	101	111
3 ¾ (45 in.)	43	53	64	74	85	95	106	116	127
4 (48 in.)	48	60	72	84	96	108	120	132	144
4 ¼ (51 in.)	55	68	82	95	109	122	136	149	163
4 ½ (54 in.)	61	76	92	107	122	137	152	167	183
4 ¾ (57 in.)	68	85	102	119	136	153	170	187	203
5 (60 in.)	75	94	113	132	150	169	188	207	225

* Increase weights by 1/3 for untamped charges. (Minimum tamping required is 10 inches.)

SECTION V. BREACHING CHARGES

86. Critical Factors and Computation

Breaching charges are applied chiefly to the destruction of concrete slab bridges, bridge piers, bridge abutments, and permanent field fortifications. The size, placement, and tamping or confinement of the breaching charge are critical factors — the size and confinement of the explosive being relatively more important because of strength and bulk of the materials to be breached. High explosive breaching charges, detonated in or against concrete and rock, must produce a shock so intense that it breaks or shatters the material. The metal reinforcing bars often found in concrete are not always cut by breaching charges. . If it is necessary to remove the reinforcement, the steel cutting formula (para (*b*) 81*b* (2)(*b*)) is used.

a. Calculation Formula. The size of a charge required to breach concrete, masonry, rock, or similar material is calculated by the formula below. By proper adjustment of the P-value, the charge size for any other explosive may be readily determined.

$P = R^3KC$

P = pounds of TNT required

R = breaching radius, in feet (*b* below)

K = material factor, given in table XI, which reflects the strength and hardness of the material to be demolished (*c* below)

C = a tamping factor, given in figure 105 which depends on the location and tamping of the charge (*d* below)

Note. For breaching walls 1 foot thick or less, increase the total calculated charge by 50 percent. Add 10 percent for charges under 50 pounds.

b. Breaching Radius R. The breaching radius

Table XI. Value of K (Material Factor) for Breaching Charges

Material	Breaching radius	K
Ordinary earth	All values	0.05
Poor masonry, shale and hardpan; good timber and earth construction	All values	0.23
Good masonry, ordinary concrete, rock	Less than 3 feet	0.35
	3 feet to less than 5 feet	.28
	5 feet to less than 7 feet	.25
	7 feet or more	.23
Dense concrete, first-class masonry	Less than 3 feet	0.45
	3 feet to less than 5 feet	.38
	5 feet to less than 7 feet	.33
	7 feet or more	.28
Reinforced concrete (concrete only; will not cut steel reinforcing)	Less than 3 feet	0.70
	3 feet to less than 5 feet	.55
	5 feet to less than 7 feet	.50
	7 feet or more	.43

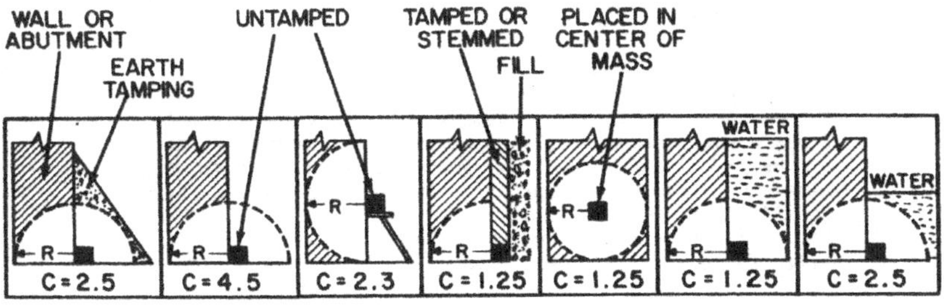

Figure 105. Value of C (tamping factor) for breaching charges.

R is the distance in feet from an explosive in which all material is displaced or destroyed. The breaching radius for external charges is the thickness of the mass to be breached. The breaching radius for internal charges is one-half the thickness of the mass to be breached if

THICKNESS OF CONCRETE IN FEET	METHODS OF PLACEMENT							DISTANCE BETWEEN CHARGES IN FEET	
	TNT (LB)						ANY		
			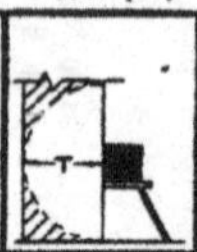					INTERNAL	EXTERNAL
COLUMN	1	2	3	4	5	6	7	8	9
2	16	28	15	8	8	16	1	2	4
2½	31	55	28	16	16	31	2	2½	5
3	41	67	38	21	21	41	4	3	6
3½	59	107	55	33	33	59	6	3½	7
4	88	159	81	49	49	88	8	4	8
4½	126	226	116	63	63	126	11	4½	9
5	157	282	144	79	79	157	16	5	10
5½	208	375	192	104	104	208	20	5½	11
6	270	486	249	135	135	270	21	6	12
6½	344	618	316	172	172	344	26	6½	13
7	369	664	340	185	185	369	33	7	14
7½	454	817	418	227	227	454	40	7½	15
8	551	991	507	276	276	551	49	8	16

Notes:

1. 10% has been added to the table for charges less than 50 lbs.
2. For best results place charge in shape of a square.
3. For thickness of concrete of 4 ft or less use charge thickness of 2 in (one block thick); over 4 ft thick, use charge thickness of 4 in (one haversack of tetrytol or plastic (M5A1)).

To use table:

1. Measure thickness of concrete.
2. Decide how you will place the charge against the concrete. Compare your method of placement with the diagrams at the top of the page. If there is any question as to which column to use, always use the column that will give you the greater amount of TNT.
3. For calculating explosives other than TNT, use relative effectiveness factor (Table VIII).

Figure 106. Breaching charge calculations.

the charge is placed midway into the mass. If holes are drilled less than halfway into the mass, the breaching radius becomes the longer distance from end of the hole to the outside of the mass. For example, if a 4-foot wall is to be breached by an internal charge placed 1 foot into the mass, the breaching radius is 3 feet. If it is to be breached by a centered internal charge, the breaching radius is 2 feet. The breaching radius is 4 feet if an external charge is used. Values of R are rounded off to the next highest ½-foot.

c. Material Factor K. K is the factor that reflects the strength and hardness of the material to be breached. Table XI gives values for the factor K for various types of material. When it is not known whether or not concrete is reinforced, it is assumed to be reinforced.

d. Tamping Factor C. The value of tamping factor C depends on the location and the tamping of the charge. Figure 105 shows typical methods for placing charges and gives values of C to be used in the breaching formula with both tamped and untamped charges. In selecting a value of C from figure 105, a charge tamped with a solid material such as sand or earth is not considered fully tamped unless it is covered to a depth equal to the breaching radius.

e. Use of Figure in Making Calculations. Figure 106 gives the weight of TNT required to breach reinforced and dense concrete targets. The weights of TNT in the table were calculated from the formula $P = R^3KC$ and the values were rounded off to the next highest pound.

f. Example. Using figure 106, calculate the amount of TNT required to breach a reinforced concrete wall 7 feet in thickness with an untamped charge placed at a distance R above the ground. From the figure (7 foot thickness and untamped charges placed at a distance R above the ground columns) the required weight of TNT is 340 pounds.

87. Placement and Number of Charges

a. Positions. In the demolition of piers and walls, the positions for the placement of explosive charges are rather limited. Unless a demolition chamber is available, the charge (or charges) may be placed against one face of the target either at ground level, somewhat above ground level, or beneath the surface. A charge placed above ground level is more effective than one placed directly on the ground. When several charges are required to destroy a pier, slab, or wall and elevated charges are desired, they are distributed equally at no less than 1 breaching radius high from the base of the object to be demolished. In this manner, the best use is obtained from the shock waves of the blast. All charges are thoroughly tamped with damp soil or filled sandbags if time permits. (Tamping must equal the breaching radius.) For piers, slabs, or walls partially submerged in water, charges are placed below the waterline. If underwater demolition is essential, the tamping factor for the placement of tamped charges with earth is used.

b. Number of Charges. The number of charges required for demolishing a pier, slab, or wall is determined by the formula:

$$N = \frac{W}{2R}$$

$N =$ number of charges

$W =$ width of pier, slab, or wall, in feet

$R =$ breaching radius in feet (para 86*b*)

If the calculated value of N has a fraction less than ½, the fraction is disregarded, but if the calculated value of N has a fraction of ½ or more, the value is "rounded off" to the next higher whole number. An exception to this general rule is in calculated N-value between 1 and 2, in which a fraction less than ¼ is disregarded, but a fraction of ¼ or more is rounded off to the next higher whole number, or 2.

Section VI. CRATERING AND DITCHING CHARGES

88. Critical Factors

a. Explosive. A special cratering charge, ammonium nitrate, issued in a waterproof metal container, is used. When the ammonium nitrate charge is not available, other explosives may be substituted.

b. Size and Placement of Charge.

(1) *Basic factors.* In deliberate cratering,

holes are bored to specific depths and spaced according to computation by formula. Deliberate craters are a minimum of 8 feet deep. In hasty cratering, holes are more shallow, contain less explosive, and are spaced on 5-foot centers. The crater depth is 1½ times the depth of the borehole. In ditching, test shots are made and the diameter and depth are increased as required. The size of the cratering charge is determined as described in paragraphs 89 and 90.

(2) *Breaching hard-surfaced pavements for cratering charges.* Hard-surfaced pavement of roads and airfields is breached so that holes may be dug for cratering charges. This is done effectively by exploding tamped charges on the pavement surface. A 1-pound charge of explosive is used for each 2 inches of pavement thickness. It is tamped with material twice as thick as the pavement. The pavement may also be breached by charges placed in boreholes drilled or blasted through it. (A shaped charge readily blasts a small-diameter borehole through the pavement and into the subgrade). Concrete should not be breached at an expansion joint, because the concrete will shatter irregularly. After breaching, holes may be made by use of the earth rod kit (para 41*c*) and widened by the detonating cord wick (para 98*b*). Also an M2A3 shaped charge (fig. 8) detonated 30 inches above any kind of soil will produce a borehole deep enough to accept a cratering charge (table XII). But since these shaped-charge boreholes are usually tapered in diameter, they should be enlarged by means of a posthole digger, detonating cord wick, or other device.

c. Confinement of Charge. Charges at cratering sites and antitank ditching sites are placed in boreholes and properly stemmed. Those at culvert sites are tamped with sandbags.

Table XII. Size of Boreholes Made by Shaped Charges

				M3 shaped charge	M2A3 shaped charge
1	Reinforced concrete	Maximum wall thickness that can be perforated		60 in	36 in
2		Depth of penetration in thick walls		60 in	30 in
3		Diameter of hole (in)	Entrance	5 in	3½ in
4			Average	3½ in	2¾ in
5			Minimum	2 in	2 in
6		Depth of hole with second charge placed over first hole		84 in	45 in
7	Armor plate	Perforation		At least 20 in	12 in
8		Average diameter of hole		2½ in	1½ in
9	Permafrost	Depth of hole with 50-in standoff		72 in	N/A
10		Depth with 30-in standoff		N/A	72 in
11		Depth with 42-in standoff		N/A	60 in
12		Diameter of hole with average (30-in) standoff		N/A	6 in to 1½ in
13		Diameter of hole with 50-in standoff		8 in to 5 in	N/A
14		Diameter of hole with normal standoff		*26–30 in to 7 in	26–30 in to 4 in
15	Ice	Depth with average (42-in) standoff		12 ft.	7 ft
16		Diameter with average (42-in) standoff		6 in	3½ in

* Boreholes made by shaped charges are cone-shaped. The diameters shown in this table are top and bottom measurements.

89. Deliberate Road Crater

A deliberate road crater may be made in all materials except loose sand, regardless of the type of road surface. The method shown in figure 107 produces a clean V-shaped crater a minimum of 8 feet deep and 25 feet wide extending about 8 feet beyond each end charge. The method of placing charges is as follows:

a. Bore the holes 5 feet apart, center-to-center, in line across the roadway. The end holes are 7 feet deep and the others are alternately 5 feet and 7 feet deep. The formula for the computation of the number of holes is:

$$N = \frac{L-16}{5} + 1$$

L = length of crater in feet measured across roadway

Any fractional number of holes is rounded off to the next highest number. If a hole is too small to accept the cratering charge, enlarge it by use of the detonating cord wick (para 98*b*).

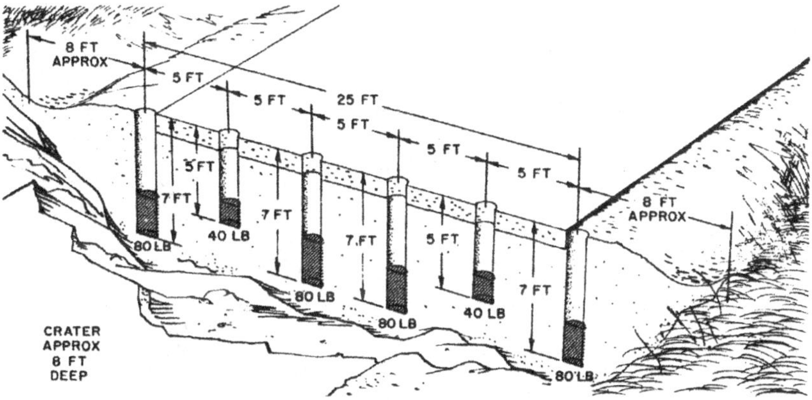

Figure 107. Charge placement for deliberate road crater.

b. Place 80 pounds of explosive in the 7-foot holes and 40 pounds of explosive in the 5-foot holes. Two 5-foot holes must not be made next to each other. If they are so calculated, one of them must be a 7-foot hole. The resulting two adjacent 7-foot holes may be placed anywhere along the line.

c. Prime all charges and connect them to fire simultaneously. A dual firing system should be used.

d. Place a 1-pound primer in each hole on top of the can for dual priming, if the standard cratering charge is used.

e. Stem all boreholes with suitable material.

90. Hasty Road Crater

Although a hasty road crater takes less time and less explosive for construction than a deliberate road crater, it is less effective because of its depth and shape. It does, however, make an excellent barrier for vehicles and small tanks (fig. 108). The method described below forms a crater about 1½ times deeper and 5 times wider than the depth of the boreholes and extends about 8 feet beyond each end charge. The sides have a slope of 30° to 60° depending on the soil. Craters formed by boreholes less than 4 feet deep and loaded with charges less than 40 pounds are ineffective against tanks. The following hasty cratering method has proved satisfactory:

a. Dig all boreholes to the same depth. This may vary from 2½ to 5 feet, depending upon the size of the crater needed. Space the holes 5 feet apart center-to-center across the road.

b. Load the boreholes with 10 pounds of explosive per foot of depth.

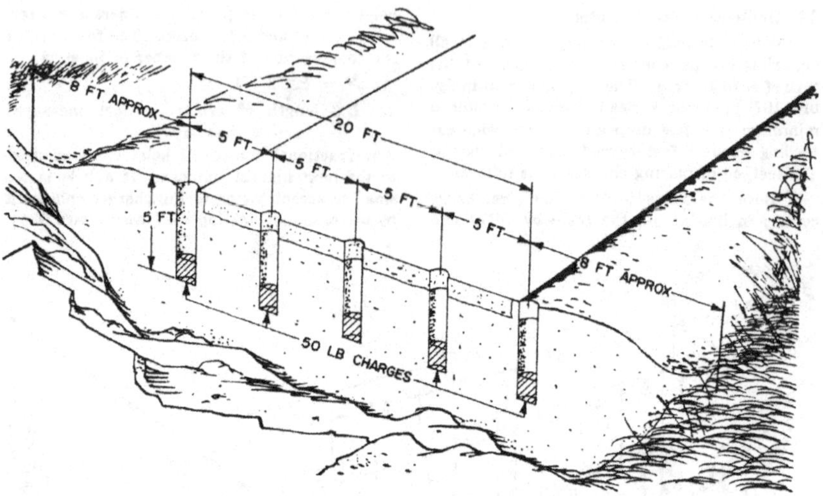

Figure 108. Charge placement for hasty road crater.

c. Prime the charges as for deliberate cratering.

d. Stem all holes with suitable material.

91. Special Cratering Methods

a. Relieved Face Cratering Method. This demolition technique produces a trapezoidal-shaped crater with unequal side slopes. The side nearest the enemy slopes at about 25° from the road surface to the bottom while that on the opposite or friendly side is about 30° to 40° steeper. The exact shape, however, depends on the type of soil found in the area of operations. In compact soil, such as clay, the relieved face cratering method will provide an obstacle shaped as shown in 1, figure 109. The procedure is as follows:

(1) Drill two rows of boreholes 8 feet apart, spacing the boreholes on 7-foot centers. On any road, the row on the friendly side will contain four boreholes. Stagger the boreholes in the row on the enemy side in relationship to the other row, as shown in 2, figure 109. This row will usually contain three boreholes or always one less than the other row (2, fig. 109).

(2) Make the boreholes on the friendly side 5 feet deep and load with 40 pounds of explosive, and those on the enemy side 4 feet deep and load with 30 pounds of explosive.

(3) Prime the charges in each row separately for simultaneous detonation. There should be a delay of detonation of ½ to 1½ seconds between rows, the row on the enemy side being detonated first. Best results will be obtained if the charges on the friendly side are fired while the earth moved in the first row is still in the air. Standard delay caps may be used for delay detonation.

(4) Acceptable results may be obtained by firing both rows simultaneously, if adequate means and sufficient time for delay firing are not available.

Note. While the procedure in (1), above, specifies four boreholes on the friendly side and three boreholes on the enemy side, more boreholes may be made if needed to crater

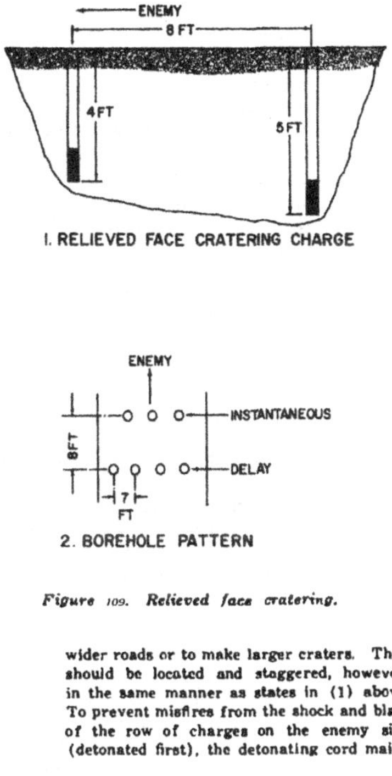

Figure 109. Relieved face cratering.

wider roads or to make larger craters. They should be located and staggered, however, in the same manner as states in (1) above. To prevent misfires from the shock and blast of the row of charges on the enemy side (detonated first), the detonating cord mains and branch lines of the row on the friendly side (detonated last) should be protected by a covering of 6 to 8 inches of earth.

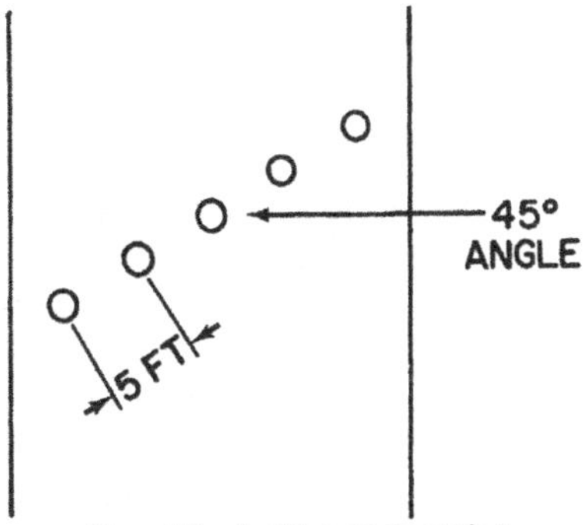

Figure 110. Angled cratering method.

b. Angled Cratering Method. This method is useful against tanks traveling in defiles or road cuts where they must approach the crater straightaway. A line of boreholes is blasted or drilled across a roadway at about a 45° angle, and charged as in figure 110. Because of the unevenness of the side slopes, tanks attempting to traverse an angled crater are usually halted effectively.

c. Standoff Distance. The standoff distance for making the boreholes on unpaved roads with M2A3 shaped charges should be 20 to 30 inches. For paved roads, the standoff distance should be about 36 inches. As the standoff distance is decreased, the depth of the open hole is decreased while the diameter is increased. In any case, test hole should be made to ascertain the optimum standoff distance.

d. Blasting Permafrost.

(1) *Number of boreholes and size of charge.* In permafrost, blasting requires about 1½ to 2 times the number of boreholes and larger charges than those calculated by standard formulas for moderate climates. Frozen soil, when blasted, breaks into large clods 12 to 18 inches thick and 6 to 8 feet in diameter. As the charge has insufficient force to blow these clods clear of the hole, they fall back into it when the blast subsides. Tests to determine the number of boreholes needed should be made before extensive blasting is attempted. In some cases, permafrost may be as difficult to blast as solid rock.

(2) *Methods of making boreholes.* Boreholes are made by three methods—standard drilling equipment, steam point drilling equipment, and shaped charges. Standard drill equipment has one serious defect—the air holes in the drill bits freeze and there is no known method of avoiding it. Steam point drilling is satisfactory in sand, silt, or clay, but not in gravel. Charges must placed immediately upon withdrawal of the steam point, otherwise the area

around the hole thaws and plugs it. Shaped charges also are satisfactory for producing boreholes, especially for cratering. Table XII shows the size of boreholes in permafrost and ice made by M3 and M2A3 shaped charges.

(3) *Explosives.* A low velocity explosive like ammonium nitrate, satisfactory for use in arctic temperatures, should be used, if available. The heaving quality of low velocity explosives will aid in clearing the hole of large boulders. If only high velocity explosives are available, charges should be tamped with water and permitted to freeze. Unless high velocity explosives are thoroughly tamped, they tend to blow out of the borehole.

e. Blasting Ice.

(1) *Access holes.* These are required for water supply and determining the thickness of ice for the computation of safe bearing pressures for aircraft and vehicles. As ice carries much winter traffic, its bearing capacity must be ascertained rapidly when forward movements are required. Small diameter access holes are made by shaped charges. On solid lake ice, the M2A3 penetrates 7 feet and the M3, 12 feet (table XII). These charges will penetrate farther but the penetration distances were tested only in ice approximately 12 feet thick. If the regular standoff is used, a large crater forms at the top, which makes considerable probing necessary to find the borehole. If a standoff of 42 inches or more is used with the M2A3 shaped charge, a clean hole without a top crater is formed. Holes made by the M2A3 average 3½ inches in diameter, while those made by the M3 average 6 inches.

(2) *Ice conditions.* In the late winter after the ice has aged, it grows weaker and changes color from blue to white. Although the structure of ice varies and its strength depends on age, air temperature, and conditions of the original formation, the same size and type of crater is formed regardless of the standoff distance. If the lake or river is not frozen to the bottom and there is a foot or more of water under the ice, the water will rise to within 6 inches of the top after the hole is blown, carrying shattered ice particles with it. This makes the hole easy to clean. If the lake is frozen to the bottom, the blown hole will fill with shattered ice and clearing will be extremely difficult. Under some conditions, shaped charges may penetrate to a depth much less than that indicated in table XII.

(3) *Surface charges.* Surface craters may be made with ammonium nitrate cratering charges or TNT, M1, M2, or M3 demolition blocks. For the best effects, the charges are placed on the surface of cleared ice and tamped on top with snow. The tendency of ice to shatter more readily than soil should be considered when charges are computed.

(4) *Underwater charges.*

(*a*) Charges are placed underwater by first making boreholes in the ice with shaped charges, and then placing the charge below the ice. An 80-pound charge of M3 demolition blocks under ice 4½ feet thick forms a crater 40 feet in diameter. This crater, however, is filled with floating ice particles, and at temperatures around 20° F freezes over in 40 minutes.

(*b*) A vehicle obstacle may be cratered in ice by sinking boreholes 9 feet apart in staggered rows. Charges (tetrytol or plastic) are suspended about 2 feet below the bottom of the ice by means of cords with sticks bridging the tops of the holes. The size of the charge depends upon the thickness of the ice. Only two or three charges are primed, usually one at each end and one at the middle. The others will detonate sympathetically. An obstacle like this may retard or halt enemy vehicles for approximately 24 hours at temperatures around −24° F.

92. Cratering at Culverts

A charge detonated to destroy a culvert not more than 15 feet deep may, at the same time, produce an effective road crater. Explosive charges should be primed for simultaneous firing and thoroughly tamped with sandbags. Culverts with 5 feet or less of fill may be destroyed by explosive charges placed in the same manner as in hasty road cratering (para 90). Concentrated charges equal to 10 pounds per foot of depth are placed in boreholes at 5-foot intervals in the fill above and alongside the culvert.

93. Antitank Ditch Cratering

a. Construction. In open country, antitank ditches are constructed to strengthen prepared defensive positions. As they are costly in time and effort, much is gained if the excavation can be made by means of cratering charges. To be effective, an antitank ditch must be wide enough and deep enough to stop an enemy tank. It may be improved by placing a log hurdle on the enemy side and the spoil on the friendly side. Ditches are improved by digging the face on the friendly side nearly vertical by means of handtools.

b. Deliberate Cratering Method. The deliberate cratering method outlined in paragraph 89 is adequate for the construction of heavy tank ditches in most types of soil.

c. Hasty Cratering Method. Ditches for medium tanks may be constructed by placing 40 pounds of cratering explosive in 4-foot holes spaced 5 feet apart. This makes a ditch approximately 6 feet deep and 20 feet wide. A heavy antitank ditch may be constructed by placing 50 pounds of cratering explosive in 5-foot holes, and spacing the holes at 5-foot intervals. The ditch will be approximately 8 feet deep and 25 feet wide (para 90).

94. Blasting of Ditches

In combat areas, ditches may be constructed to drain terrain flooded by the enemy or as initial excavations for the preparation of entrenchments. Rough open ditches 2½ to 12 feet deep and 4 to 40 feet wide may be blasted in most types of soils. A brief outline of the method is given below.

a. Test Shots. Before attempting the actual ditching, make test shots to determine the proper depth, spacing, and weight of charges needed to obtain the required results. Make beginning test shots with holes 2 feet deep and 18 inches apart and then increase the size of the charge and the depth as required. A rule of thumb for ditching is to use 1 pound of explosive per cubic yard of earth in average soil.

b. Alinement and Grade. Mark the ditch centerline by transit line or expedient means and drill holes along it. When a transit or hand level is used, the grade of the ditch may be accurately controlled by checking the hole depth every 5 to 10 holes and at each change in grade. In soft ground, the holes may be drilled with a miner's drill or earth auger. Holes are loaded and tamped immediately to prevent cave-ins and insure that the charges are at proper depth. Ditches are sloped at a rate of 6 to 12 inches per 100 feet.

c. Methods of Loading and Firing.

(1) *Propagation method.* By this method only one charge is primed—the charge placed in the hole at one end of the line of holes made to blast the ditch. The concussion from this charge sympathetically detonates the next charge and so on until all are detonated. Only commercial dynamite should be used in this operation. The propagation method is effective, however, only in moist soils or in swamps where the ground is covered by several inches of water. If more than one line of charges is required to obtain a wide ditch, the first charge of each line is primed. The primed hole is overcharged 1 or 2 pounds.

(2) *Electrical method.* Any high explosive may be used in ditching by the electrical firing method which is effective in all soils except sand, regardless of moisture content. Each charge is primed with an electric cap and the caps are connected in series. All charges are fired simultaneously.

(3) *Detonating cord method.* In this ditching method any high explosive may be used. It is effective in any type of soil, except sand, regardless of moisture content. Each charge is primed with detonating cord and connected to a main or ring main line.

Section VII. LAND CLEARING AND QUARRYING CHARGES

95. Introduction

In military operations, construction jobs occur in which demolitions may be employed to advantage. Among these jobs are land clearing, which includes stump and boulder removal, and quarrying. The explosives commonly used are military dynamite and detonating cord. The quantity of explosive used is generally calculated by rule of thumb. Charges may be placed in boreholes in the ground under or at the side of the target, in the target itself, or on top of the target. All charges should be tamped or mudcapped, which is a form of light tamping.

96. Stump Removal

In certain military operations it may be necessary to remove stumps as well as trees. Stumps are of two general types, tap- and lateral-rooted (fig. 111). Military dynamite is the explosive best suited for stump removal. A rule of thumb is to use 1 pound per foot of diameter for dead stumps and 2 pounds per foot for live stumps, and if both tree and stump are to be removed, to increase the amount of explosive by 50 percent. Measurements are taken at points 12 to 18 inches above the ground.

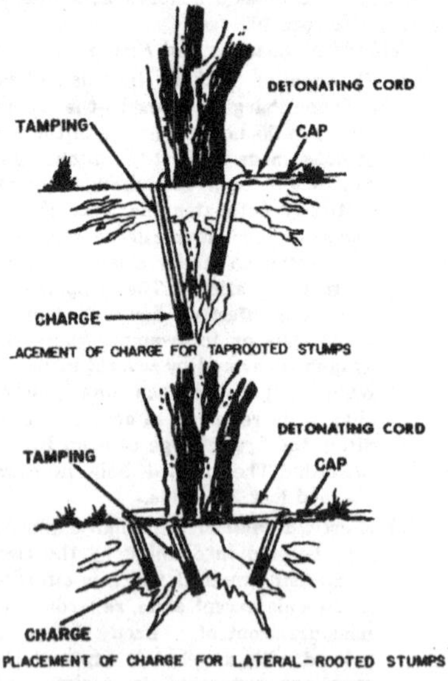

Figure 111. Stump blasting charges.

a. Taproot Stumps. For taproot stumps, one method is to bore a hole in the taproot below the level of the ground. The best method is to place charges on both sides of the taproot to obtain a shearing effect (fig. 111). For best results, tamp the charges.

b. Lateral-Root Stumps. In blasting lateral root stumps, drill sloping holes as shown in figure 111. Place the charge as nearly as possible under the center of the stump and at a depth approximately equal to the radiús of the stump base. If for some reason the root formation cannot be determined, assume that it is the lateral type and proceed accordingly.

97. Boulder Removal

In the building of roads and airfields or other military construction, boulders can be removed by blasting. The most practical methods are snakeholing, mudcapping, and blockholing.

a. Snakeholing Method. By this method, a hole large enough to hold the charge is dug under the boulder. The explosive charge is packed under and against the boulder as shown in figure 112. For charge size, see table XIII.

b. Mudcapping Method. For surface or slightly embedded boulders, the mudcapping method is very effective. The charge is placed on top or against the side of the boulder and covered with 10 to 12 inches of mud or clay (fig. 112). For charge size see table XIII.

c. Blockholing Method. This method is very effective on boulders lying on the surface or slightly embedded in the earth. A hole is drilled

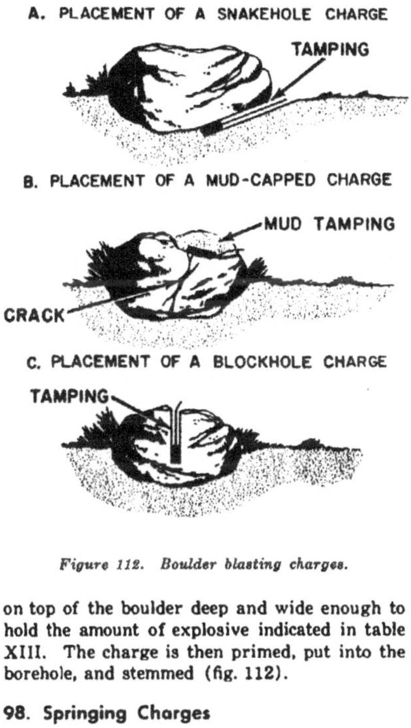

Figure 112. Boulder blasting charges.

on top of the boulder deep and wide enough to hold the amount of explosive indicated in table XIII. The charge is then primed, put into the borehole, and stemmed (fig. 112).

98. Springing Charges

a. Definition and Method. A springing charge is a comparatively small charge detonated in the bottom of a drilled borehole to form an enlarged chamber for placing a larger charge. At times two or more springing charges in succession may be needed to make the chamber large enough for the final charge. Under these conditions at least 30 minutes should be allowed between firing and placing successive charges for the boreholes to cool.

Table XIII. Charge Size for Blasting Boulders

Boulder diameter (ft)	Pounds of explosive required		
	Blockholing	Snakeholing	Mudcapping
1½ (rare)	⅛	½	1
2	⅛	½	1½
3	¼	¾	2
4	⅜	2	3½
5	½	3	6

b. Detonating Cord Wick. This is several strands of detonating cord taped together and used to enlarge boreholes in soils. One strand generally widens the diameter of the hole about 1 inch.

(1) A hole is made by driving a steel rod approximately 2 inches in diameter into the ground to the depth required (para 41*c*) or by means of a shaped charge. According to the rule of thumb, a hole 10 inches in diameter requires 10 strands of detonating cord. These must extend the full length of the hole and be taped or tied together into a "wick" to give optimum results. The wick may be placed into the hole by an inserting rod or some field expedient. Firing may be done electrically or nonelectrically. An unlimited number of wicks may be fired at one time by connecting them by a detonating cord ring main or main line.

(2) The best results from the use of the detonating cord wick are obtained in hard soil. If successive charges are placed in the holes, excess gases must be burned out and the hole inspected for excessive heat.

99. Quarrying

Quarrying is the extraction of rock in the natural state. Military quarries, generally of the open face type, are developed by the single or multiple bench method. See TM 5–332 for detailed information.

CHAPTER 4

DEMOLITION PROJECTS

Section I. DEMOLITION PLAN

100. Scope

Thus far, this manual has been concerned with methods and techniques in the selection, calculation, priming, placement, and firing of explosives on such materials as steel, concrete, wood, and stone and in earth. This chapter deals with the problems of applying these techniques to the conduct of demolitions projects.

101. Reconnaissance to Develop Demolition Plan

a. Information Required. Thorough reconnaissance is necessary before an effective plan may be made to demolish a target, as reconnaissance provides information in all areas related to the project. Before the demolition of bridges, culverts, and road craters, the following data is provided by reconnaissance.

(1) Situation map sketch (fig. 113) showing the relative position of the objects to be demolished, the surrounding terrain features, and the coordinates of the objects keyed to existing maps.

(2) Side-view sketch of the demolition object. If, for example a bridge is to be blown, a sketch showing the overall dimensions of critical members is necessary (fig. 114).

(3) Cross section sketches, with relatively accurate dimensions of each member to be cut (fig. 114).

(4) A bill of explosives, showing the quantity and kind required.

(5) Sketch of the firing circuits.

(6) List of all equipment required for the demolition.

(7) List of all unusual features of the site.

(8) Estimate of time and labor required to bypass the site.

(9) Estimate of time and labor required for the demolition.

(10) Estimate and sketch of security details required.

SITUATION MAP SKETCH (INCLUDES PRINCIPAL TERRAIN FEATURES; IMMEDIATE AVENUES OF APPROACH; OBSERVATION AND COVER; MAP COORDINATES)

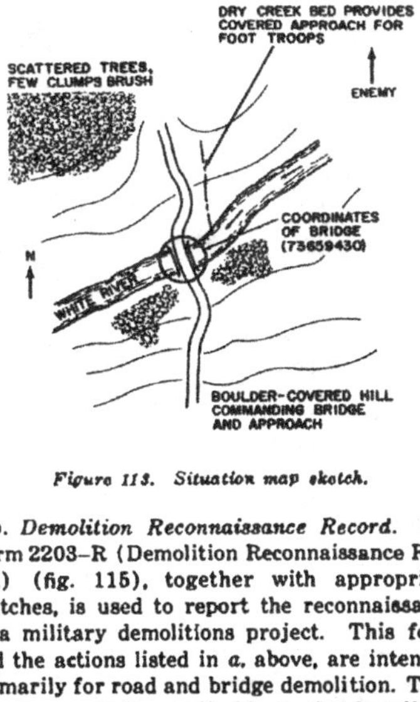

Figure 113. Situation map sketch.

b. Demolition Reconnaissance Record. DA Form 2203–R (Demolition Reconnaissance Record) (fig. 115), together with appropriate sketches, is used to report the reconnaissance of a military demolitions project. This form and the actions listed in *a*, above, are intended primarily for road and bridge demolition. They are also partially applicable to the demolition of almost any other object. In certain instances, the report may require a security classification. The form is reproduced locally on 8- by 10½-inch paper.

102. Demolition Orders

a. Purpose. Three commanders are usually involved in the execution of a demolition proj-

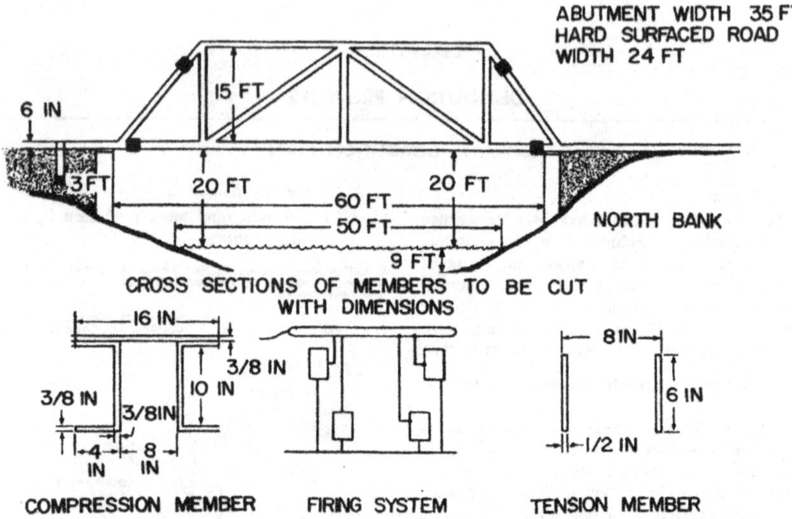

Figure 114. Drawing of object to be demolished.

ect. These are the tactical commander with over-all responsibility and authority to order the firing of the demolition, the commander of the demolition guard, and the commander of the demolition firing party. To assist the commanders in the execution of their responsibilities, two demolition orders are used. These are shown in figure 116, (Orders to the Demolition Guard Commander) and in figure 117, (Orders to the Commander, Demolition Firing Party) (DA Form 2050-R). The procedures that follow are in accord with the agreement between the armed forces of NATO nations and will be complied with by Department of the Army units.

b. Procedures. Each authorized commander, or the tactical commander referred to in *a* above, will—

(1) Establish the requirement and assign the responsibility for a demolition guard and a demolition firing party.

(2) Establish a clear cut channel whereby the order to fire the demolition is transmitted from himself to the commander of the demolition guard and thereby to the commander of the demolition firing party. In the event that no demolition guard is required, this channel must be established between the authorized commander and the commander of the demolition firing party.

(3) Insure that this channel is known and understood by all concerned, and that positive and secure means of transmitting the order to fire are established.

(4) Specify the conditions for executing the demolition as contained in part V of "Orders to the Commander, Demolition Firing Party," and completing part IV of the "Orders to the Demolition Guard Commander."

DEMOLITION RECONNAISSANCE RECORD
(FM 5-25)

SECTION I - GENERAL

1. FILE NO.	2. DML RECON REPORT NO.	3. DATE	4. TIME
411	1	21 June 63	2100

5.	NAME	GRADE	ORGANIZATION
RECON ORDERED BY	John J. Shuba	Capt.	954 Engr Bn
6. PARTY LEADER	Joe C. Gay	SFC	Recon Sec

7. MAP REFERENCE Stanton 1:25,000 Sheet No 6578 IV SE

8. SITE AND OBJECT	9. TIME OBSERVED	10. LOCATION
Steel truss bridge across White River on Hwy 1	2200	73659430

11. GENERAL DESCRIPTION Single span steel truss construction. Length of span 60 ft. Width of bridge 24 ft. Height of truss 15 ft. Height of bridge above water 20 ft. Concrete abutments 35 ft wide.

12. NATURE OF PROPOSED DEMOLITION Rotation method. Cut upper and lower chord on both ends of truss on the upstream side of the bridge. Destroy south shore abutment.

SECTION II - ESTIMATES*

13. EXPLOSIVES REQUIRED

a. TYPES	b. POUNDS	c. CAPS Electric	c. CAPS Nonelectric	d. DETONATING CORD	e. FUSE LIGHTERS
TNT Cratering chg	20 lbs 6-40 lb chgs	4	0	500 ft	0

14. EQUIPMENT REQUIRED

1- 2½ ton truck 7 extra reels of firing wire.

1- Squad demo set.

15. PERSONNEL AND TIME REQUIRED	PERSONNEL	TIME
	1 squad	1 hr

SECTION III - REMARKS

16. UNUSUAL FEATURES OF SITE

None

17. LABOR AND TIME ESTIMATED REQUIRED FOR BYPASS

*Determine availability of Items 13, 14, and 15 before reconnaissance.

DA FORM 2203-R, 1 Mar 67 Previous edition of this form is obsolete.

Figure 115. Demolition reconnaissance record.

CLASSIFICATION

Serial No.__________________Security Classification__________

ORDERS TO THE DEMOLITION GUARD COMMANDER

Notes:
1. This form will be completed and signed before it is handed to the Commander of the Demolition Guards.
2. In completing the form, all spaces must either be filled in or lined out.
3. The officer empowered to order the firing of the demolition is referred to throughout as the "Authorized Commander".

From__________________________To______________________________

PART I - PRELIMINARY INSTRUCTIONS

1.a. Description of target______________________________________

b. Location:

Map Name and Scale________________Sheet No._______________

Grid Reference__________________________

c. Codeword or codesign (if any) of demolition target.________

2. The Authorized Commander is_____________________________________ (give appointment only). If this officer should delegate his authority, you will be notified by one of the methods shown in paragraph 4, below.

3. The DEMOLITION FIRING PARTY has been/will be provided by______ _______________

4. All messages, including any codewords or codesign (if any) used in these orders, will be passed to you by:

a. Normal command wireless net, or

b. Special liaison officer with communications direct to the Authorized Commander, or

CLASSIFICATION

Figure 116. Orders to the Demolition Guard Commander. ①

CLASSIFICATION (Cont'd)

c. Telephone by the Authorized Commander, or

d. The Authorized Commander personally, or

e. ____

(Delete those NOT applicable)

Note: All orders sent by message will be prefixed by the codeword or codesign (if any) at paragraph 1.c., and all such messages must be acknowledged.

CLASSIFICATION (Cont'd)

Figure 116—Continued. ②

NATO - UNCLASSIFIED

PART II - CHANGING STATES OF READINESS

5. The demolition will be prepared initially to the State of Readiness______________by__________hours on_________(date).

6. On arrival at the demolition site, you will ascertain from the commander of the demolition firing party the estimated time required to change from State "1" (SAFE) to State "2" (ARMED). You will ensure that this information is passed to the Authorized Commander and is acknowledged.

7. Changes in the State of Readiness from State "1" (SAFE) to State "2" (ARMED) or from State "2" to State "1" will be made only when so ordered by the Authorized Commander. However, the demolition may be ARMED in order to accomplish emergency firing when you are authorized to fire it on your own initiative.

8. A record of the changes in the State of Readiness will be entered by you in the table below, and on the firing orders in possession of the commander of the demolition firing party.

State of Readiness ordered "1" (SAFE) or "2" (ARMED)	Time & date change to be completed	Authority	Time & date of receipt of order

Note: If the order is transmitted by an officer in person, his signature and designation will be obtained in the column headed "Authority".

9. You will report completion of all changes in the State of Readiness to the Authorized Commander by the quickest means.

PART III - ORDERS FOR FIRING THE DEMOLITION

10. The order for firing the demolition will be passed to you by the Authorized Commander.

NATO - UNCLASSIFIED

Figure 116—Continued. ⑧

NATO - UNCLASSIFIED (Cont'd)

PART III - ORDERS FOR FIRING THE DEMOLITION

11. On receipt of this order you will immediately pass it to the commander of the demolition firing party on his demolition Orders form ("Orders to the Commander of the Demolition Firing Party").

12. After the demolition has been fired you will report the results immediately to the Authorized Commander.

13. In the event of a misfire or only partially successful demolition you will give the firing party protection until such time as it has completed the demolition and report again after it has been completed.

NATO - UNCLASSIFIED (Cont'd)

Figure 116—Continued. ④

NATO - UNCLASSIFIED

PART IV - EMERGENCY FIRING ORDERS

Notes: 1. One sub-paragraph of paragraph 14 must be deleted.

2. The order given herein can only be altered by the issue of a new form, or, in emergency by the appropriate order (or codeword if used) in Part V.

14.a. You will order the firing of the demolition only upon the order of the Authorized Commander.

OR

b. If the enemy is in the act of capturing the target, and/or munition, you will order the firing of the demolition on your own initiative.

PART V - CODEWORDS (IF USED)

	Action to be taken	Codeword (if used)
a.	Change State of Readiness from "1" to "2" (See paragraph 7)	
b.	Change State of Readiness from "2" to "1" (See paragraph 7)	
c.	Fire The Demolition (see paragraph 10)	
d.	Paragraph 14a is now cancelled. You are now authorized to fire the demolition if the enemy is in the act of capturing it.	

NATO - UNCLASSIFIED

Figure 116—Continued. ⑤

NATO - UNCLASSIFIED (Cont'd)

PART IV - EMERGENCY FIRING ORDERS

e.	Paragraph 14b is now cancelled. You will order the firing of the demolition only upon the order of the Authorized Commander.	Codeword (if used)
f.	Special authentication instructions, if any.	

PART VI

Signature of officer issuing these orders__________

Name (printed in capital letters)__________

Rank__________Appointment__________

Time of issue______hours,__________(date).

NATO - UNCLASSIFIED (Cont'd)

Figure 116—Continued ⑥

NATO - UNCLASSIFIED

PART VII - DUTIES OF THE COMMANDER

OF THE DEMOLITION GUARD

15. You are responsible for:-

a. Command of the demolition guard and the demolition firing party.

b. The safety of the demolition from enemy attack, capture, or sabotage.

c. Control of traffic and refugees.

d. Giving the orders to the demolition firing party in writing to change the state of readiness.

e. Giving the order to the demolition firing party in writing to fire the demolition.

f. After the demolition, reporting on its effectiveness to the Authorized Commander.

g. Keeping the Authorized Commander informed of the operational situation at the demolition site.

16. You will acquaint yourself with the orders issued to the Commander of the Demolition Firing Party and with the instructions given by him.

17. The Demolition Guard will be so disposed as to ensure at all time complete all-round protection of the demolition against all types of attack or threat.

18. The Commander of the Demolition Firing Party is in technical control of the demolition. You will agree with him on the site of your Headquarters and of the firing point. These should be together whenever practicable. When siting them you must give weight to the technical requirements of being able to view the demolition and have good access to it from the firing point.

NATO- UNCLASSIFIED

Figure 116—Continued. ①

NATO - UNCLASSIFIED (Cont'd)

PART VII - DUTIES OF THE COMMANDER

OF THE DEMOLITION GUARD

19. You will nominate your deputy forthwith and compile a seniority roster. You will ensure that each man knows his place in the roster, understands his duties and knows where to find this form if you become a casualty or are unavoidably absent. The seniority roster must be made known to the Commander of the Demolition Firing Party.

20. Once the state of readiness "2 ARMED" has been ordered, either you or your deputy must always be at your Headquarters so that orders can be passed on immediately to the Commander of the Demolition Firing Party.

NATO - UNCLASSIFIED (Cont'd)

Figure 118—Continued. ®

SECURITY CLASSIFICATION

ORDERS TO THE COMMANDER, DEMOLITION FIRING PARTY	SERIAL NUMBER

NOTE: Parts I, II and III will be completed and signed before this form is handed to the commander of the Demolition Firing Party. Paragraphs 4 and 5 can only be altered by the authority issuing these orders. In such cases a new form will be issued and the old one destroyed.

FROM:	TO:

PART I - ORDERS FOR PREPARING AND CHARGING THE DEMOLITION TARGET

1a. DESCRIPTION

b. LOCATION			c. CODE WORD OF DEMOLITION TARGET (If any)
MAP NAME AND SCALE	SHEET NO.	GRID REFERENCE	

d. ATTACHED PHOTOGRAPHS AND SPECIAL TECHNICAL INSTRUCTIONS

2. THE DEMOLITION GUARD IS BEING PROVIDED BY (Unit)

3. YOU WILL PREPARE AND CHARGE THE DEMOLITION TARGET TO THE STATE OF READINESS ____________ BY ____________ HOURS ON (Date) ____________
ANY CHANGES MAY ONLY BE MADE ON THE ORDER OF THE ISSUING AUTHORITY, OR BY THE OFFICER DESIGNATED IN PARAGRAPH 4d AND WILL BE RECORDED BELOW.

STATE OF READINESS ORDERED "1(SAFE)" or "2(ARMED)"	TIME AND DATE CHANGE TO BE COMPLETED	AUTHORITY	TIME AND DATE OF RECEIPT OF ORDER

NOTE: All orders received by message will be verified by the code word at Paragraph 1c. If the order is transmitted by an officer in person, his signature and designation will be obtained in the Column headed "Authority".

PART II - ORDERS FOR FIRING

NOTE: The officer issuing these orders will strike out the subparagraphs of Paragraphs 4 and 5 which are not applicable. When there is a demolition guard, Paragraph 4 will always be used and Paragraph 5 will always be struck out.

4a. YOU WILL FIRE THE DEMOLITION AS SOON AS YOU HAVE PREPARED IT.
b. YOU WILL FIRE THE DEMOLITION AT ____________ HOURS ON (Date) ____________.
c. YOU WILL FIRE THE DEMOLITION ON RECEIPT OF THE CODE WORD ____________.
d. YOU WILL FIRE THE DEMOLITION WHEN THE OFFICER WHOSE DESIGNATION IS ____________ ____________ HAS SIGNED PARAGRAPH 8 BELOW.

EMERGENCY FIRING ORDERS (ONLY applicable when there is NO demolition guard)

5a. YOU WILL NOT FIRE THE DEMOLITION IN ANY CIRCUMSTANCES EXCEPT AS ORDERED IN PARAGRAPH 4 ABOVE.
b. YOU WILL FIRE THE DEMOLITION ON YOUR OWN INITIATIVE IF THE ENEMY IS IN THE ACT OF CAPTURING IT.

DA FORM 2050-R, 1 NOV 57 SECURITY CLASSIFICATION

Figure 117. Orders to the Commander, Demolition Firing Party. ①

SECURITY CLASSIFICATION

PART III - ORDERS FOR REPORTING			
6. AFTER FIRING THE DEMOLITION YOU WILL IMMEDIATELY REPORT RESULTS TO THE OFFICER WHO ORDERED YOU TO FIRE. IN THE EVENT OF A PARTIAL FAILURE YOU WILL WARN HIM, AND IMMEDIATELY CARRY OUT THE WORK NECESSARY TO COMPLETE THE DEMOLITION			
7. FINALLY, YOU WILL IMMEDIATELY REPORT THE RESULTS TO YOUR UNIT COMMANDING OFFICER (See Paragraph 13.)			
SIGNATURE OF OFFICER ISSUING THESE ORDERS	NAME (In capitals) DESIGNATION	TIME OF ISSUE	DATE OF ISSUE
PART IV - ORDER TO FIRE			
8. BEING EMPOWERED TO DO SO, I ORDER YOU TO FIRE NOW THE DEMOLITION DESCRIBED IN PARAGRAPH 1.			
SIGNATURE	NAME (In capitals) DESIGNATION	TIME	DATE

PART V - GENERAL INSTRUCTIONS (Read These Instructions Carefully)

9. YOU ARE IN TECHNICAL CHARGE OF THE PREPARATION, CHARGING AND FIRING OF THE DEMOLITION TARGET DESCRIBED. YOU WILL NOMINATE YOUR DEPUTY FORTHWITH AND COMPILE A SENIORITY ROSTER OF YOUR PARTY. YOU WILL INSURE THAT EACH MAN KNOWS HIS PLACE IN THE ROSTER, UNDERSTANDS THESE INSTRUCTIONS, AND KNOWS WHERE TO FIND THIS FORM IF YOU ARE HIT OR UNAVOIDABLY ABSENT. YOU WILL CONSULT WITH THE COMMANDER OF THE DEMOLITION GUARD ON THE SITING OF THE FIRING POINT.

10. YOU MUST UNDERSTAND THAT THE COMMANDER OF THE DEMOLITION GUARD (where there is one) IS RESPONSIBLE FOR:
   a. OPERATIONAL COMMAND OF ALL THE TROOPS AT THE DEMOLITION SITE. (You are therefore under his command.)
   b. PREVENTING THE CAPTURE OF THE DEMOLITION SITE, OR INTERFERENCE BY THE ENEMY WITH DEMOLITION PREPARATIONS.
   c. CONTROLLING ALL TRAFFIC AND REFUGEES.
   d GIVING YOU THE ORDER TO CHANGE THE STATE OF READINESS FROM "1(SAFE)" TO "2(ARMED)" OR BACK TO "1(SAFE)" AGAIN. YOU WILL INFORM HIM OF THE TIME REQUIRED FOR SUCH A CHANGE.
   e. PASSING TO YOU THE ACTUAL ORDER TO FIRE.

11. WHEN THERE IS NO DEMOLITION GUARD AND YOU ARE INSTRUCTED IN PARAGRAPH 4 TO ACCEPT THE ORDER TO FIRE FROM SOME PARTICULAR OFFICER, IT IS IMPORTANT THAT YOU ARE ABLE TO IDENTIFY HIM.

12. IF YOU GET ORDERS TO FIRE, OTHER THAN THOSE LAID DOWN IN PARAGRAPH 4, YOU SHOULD REFER THEM TO THE DEMOLITION GUARD COMMANDER OR, IF THERE IS NO DEMOLITION GUARD COMMANDER, TO YOUR IMMEDIATE SUPERIOR. IF YOU CANNOT DO THIS, YOU WILL ONLY DEPART FROM YOUR WRITTEN INSTRUCTIONS WHEN YOU ARE SATISFIED AS TO THE IDENTITY AND OVERRIDING AUTHORITY OF WHOEVER GIVES YOU THESE NEW ORDERS, AND YOU WILL GET HIS SIGNATURE IN PARAGRAPH 8 WHENEVER POSSIBLE.

13. THE REPORT TO YOUR UNIT COMMANDING OFFICER, AS CALLED FOR IN PARAGRAPH 7, SHOULD CONTAIN THE FOLLOWING INFORMATION (where applicable):
   a. IDENTIFICATION REFERENCE OF DEMOLITION.
   b. MAP REFERENCE.
   c. TIME AND DATE WHEN DEMOLITION WAS FIRED.
   d. EXTENT OF DAMAGE ACCOMPLISHED, INCLUDING:
      ESTIMATED WIDTH OF GAP) IN CASE OF A BRIDGE
      NUMBER OF SPANS DOWN)
      SIZE AND LOCATION OF CRATERS IN A LOAD OR RUNWAY.
      MINES LAID.
   e. SKETCH SHOWING EFFECT OF DEMOLITION.

SECURITY CLASSIFICATION

Figure 117--Continued. ①

c. Orders to the Demolition Guard Commander. The authorized commander completes and signs this form. The order is written in seven parts, each of which is self-explanatory.

d. Orders to the Commander, Demolition Firing Party. In addition to those items listed in *b* above, the authorized commander designates the unit or individual responsible for the preparation of these orders. This unit or individual will complete and sign parts I through III and pass the order to the commander of the demolition firing party. Part IV will be completed upon detonation of the demolition (fig. 117).

e. Definitions. The states of readiness (*safe and armed*) referred to in part I of the Order to the Commander, Demolition Firing Party, and in part II of the Order to the Demolition Guard Commander, are described as follows:

(1) "1 (*Safe*)." The explosive charges are prepared and securely fixed to the target and are safe against premature firing. All firing circuits and accessories have been checked, are in proper operating condition, and are ready to be attached to charges. If detonating cord is used it may be attached to demolition charges; however, detonators will not be attached to detonating cord ring mains or main lines until the state of readiness is changed to "armed."

(2) "2 (*Armed*)." The demolition is ready for immediate firing. The risk of premature detonation is accepted.

f. Disposition of Orders. After the demolition has been fired, one copy of the orders will be retained by the headquarters of the issuing authority and one by the commander of the demolition firing party.

Section II. TECHNIQUES COMMON TO MOST DEMOLITIONS

103. Types of Military Demolitions

There are three types of demolitions applicable to tactical situations—reserved, deliberate, and hasty.

a. Reserved Demolitions. These are specifically controlled at a command level appropriate to the tactical or strategic plan. Reserved demolitions are usually in place, "ready and waiting," in the "safe" condition.

b. Deliberate Demolitions. Deliberate demolitions are used when enemy interference during preparations is unlikely and there is sufficient time for thorough reconnaissance and careful preparation. Deliberate preparation permits economy in the use of explosives, since time permits accurate calculation and positive charge placement to obtain the effects required.

c. Hasty Demolitions. Hasty demolitions are used when time is limited and economy of explosives is secondary to speed. In all cases, common sense and good judgment must be exercised to prevent waste. In the preparation of demolition projects in forward areas where a surprise raid by hostile forces is possible, a priority should be given to each charge. Although this procedure is relatively time consuming, it causes maximum damage to the project in relation to the time required, even though enemy interference might prevent completion of the job. Each charge is primed as it is placed; for if charges are all placed first and then primed, it is possible that enemy interference prior to the act of priming might stop the work before any damage is done. The use of dual detonating cord ring main lines and branch lines is recommended for all frontline demolition projects (para 64–70).

104. Nuclear Weapons Demolitions

Atomic demolition munitions (ADM) may be effectively employed to create obstacles and to destroy and deny military facilities or installations. They have the capability of creating large radioactive craters with little preparatory effort. The residual radiation and fallout hazards require consideration; however, the use of small yields minimizes the fallout hazard and area of residual contamination. The ODM, like conventional hand-placed charges, has a primary advantage of no delivery error, which permits the use of minimum yield for a given target. This is of particular importance in producing craters or for destruction through cratering effects since the radius of cratering effects of atomic weapons is relatively small in comparison with other effects. No further information, see FM 5–26.

105. Supplementing Demolition Obstacles

Nuisance mining and charges with delay fuses are a very potent means of increasing the effects of demolition projects. The area to be mined should include the facility to be destroyed, the ground where a replacement structure or remedial work will likely be performed, working party bivouacs, and alternate sites. Thus, for a demolished bridge, the dropped spans and abutments should be mined to impede removal or recovery; suitable sites for a floating bridge or ford should be mined to prevent ready use; and locations likely to be selected for material storage, equipment parks, or bridge unit bivouacs should also be well mined and boobytrapped.

Section III. BRIDGE DEMOLITION

106. Extent of Demolition

There is no rule of thumb or regulation to indicate the optimum extent of demolition of bridges. It is determined after investigation and analysis of specific conditions.

a. Complete Bridge Demolition. Complete demolition leaves nothing of the old bridge suitable for use in a new bridge. Debris is left on the site where its removal will require much hazardous work before any kind of crossing can be installed. However, when enough demolition is accomplished to force the enemy to select another site for a temporary bridge as a substitute for the damaged bridge, further demolition is unnecessary. Too, a permanent structure is not likely to be replaced in kind during wartime. However, where the terrain is such that the existing bridge site is needed for a new structure, even a temporary one, demolition in greater proportions may be justified.

b. Partial Demolition.

(1) *Method.* Bridges are generally demolished to create obstacles that delay the enemy. This seldom requires complete destruction. Unless a denial operation is in effect, the demolition method chosen should permit the economical reconstruction of the bridge by friendly troops at a later date. Frequently the necessary delay can be obtained by only blasting a gap too long to be spanned by the prefabricated bridging available to the enemy. This gap should be located where the construction of an intermediate support is difficult or impossible. A high and relatively slender bridge component may be demolished by cutting one side so that it topples into a mass of broken and twisted material. The destruction of massive bridge components, however, requires large expenditures of explosive, time, equipment, and effort that may not be profitable. In many cases on major bridges, the destruction of any component that can easily be replaced may not be justified.

(2) *Factors determining the extent of destruction.* Factors that determine the extent of destruction needed for a project are as follows:

(*a*) The tactical and strategical situations that indicate the length of time the enemy must be delayed, the time available for demolition, and the extent of denial to be accomplished.

(*b*) The likelihood that friendly forces may reoccupy the area and require the use of the bridge.

(*c*) The results to be obtained by the expenditure of labor and materials compared with the results that may be obtained elsewhere with the same effort.

(*d*) The manpower, equipment, and kinds and quantities of explosives available.

107. Parts of Fixed Bridges

The ordinary fixed bridge is divided into two main parts: the lower part or substructure, and the upper part or superstructure (fig. 118).

a. Substructure. The substructure consists of the parts of the bridge that support the superstructure. There are two kinds of supports: end supports or abutments and intermediate supports, or piers or bents. The parts of the substructure are—

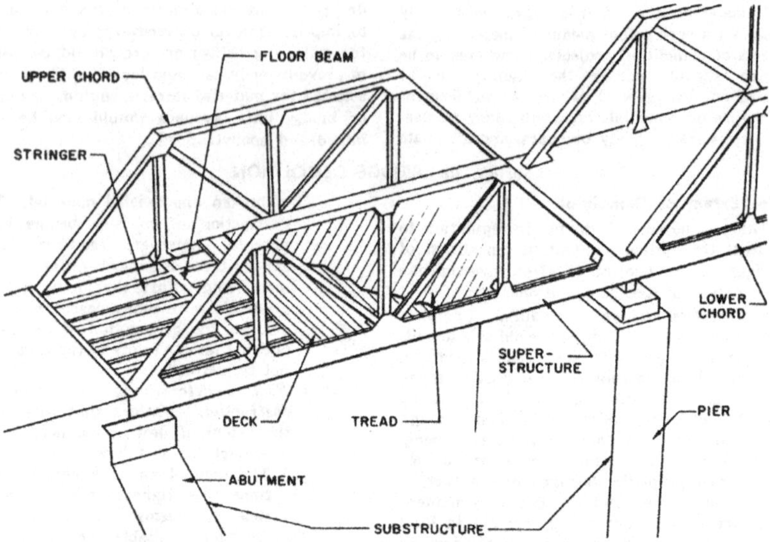

Figure 118. Parts of fixed bridge.

(1) *Abutment.* The ground supports at the ends of a bridge are called abutments. These may be constructed of concrete, masonry, steel, or timber and may include retaining walls or an end dam.

(2) *Footing.* A footing is that part of any bridge support that rests directly on the ground. It distributes the load over an area wide enough to keep the support from sinking into the ground.

(3) *End dam.* An end dam is a retaining wall of concrete, wood, or other material at the end of a bridge that supports the bank and keeps the approach road from caving in.

(4) *Intermediate support.* An intermediate support is a support placed beneath a bridge between the abutments. It may be a pier of masonry or concrete, cribbing, several pile or trestle bents constructed as a unit, or a single pile or trestle bent.

b. Superstructure. The superstructure includes the flooring, stringers, floor beams, and any girders or trusses that make up the total part of the bridge above the substructure (fig. 118).

(1) *Span.*

(*a*) *Simple.* Simple spans have stringers that extend only from one support to the next.

(*b*) *Continuous.* Continuous spans have beams that extend over one or more intermediate supports.

(2) *Truss.* A truss consists of these principal elements:

(*a*) *Lower chord.* The lower chord is the lower member in a panel of a truss that runs parallel to the deck.

(*b*) *Upper chord.* The upper chord includes the upper members in the panel.

(3) *Stringers.* Stringers run longitudinally with the bridge and directly support the deck.

(4) *Deck and tread.* The deck is the floor of the bridge and the tread, the top surface material.

108. Planning Bridge Demolitions

a. Structural Characteristics. The demolition of bridges must be carefully planned, as bridges have a great variety of superstructures made of steel, timber, or masonry and various types of substructures made of these materials. The size and placement of the charge, therefore, depends on the characteristics of the individual bridge structure.

b. General Procedures. Some general procedures apply to most bridge demolition projects; *for example:* if charges are placed under the bridge roadway, special precautions must be taken to insure that the charges will not be shaken loose or initiated by traffic on the bridge. The following general points apply to the demolition of most or all of the bridge structures mentioned and described below.

(1) Hasty charges, which must be placed first because of enemy interruption, should be located carefully, if possible, so that they may be included later on into the deliberate preparation of the bridge.

(2) It is often possible either to economize on the use of explosives or to improve the thoroughness of the demolition by blasting several times rather than only once. When conditions permit, this procedure should be considered.

(3) Tension members are more difficult to repair than compression members, because the latter may sometimes be replaced by cribbing while the former almost always require steel riveting or welding. Thus tension members should be given priority.

(4) When bridges over railways or canals are to be destroyed, the demolition should be so planned that any temporary intermediate piers that might be erected to repair the structure will be located where they will block traffic on the railroad or canal.

(5) Any long steel members that require cutting in only one place to demolish the bridge should be further damaged to prevent their ready salvage by recutting or splicing. It is not necessary to cut such members completely in two at other points to accomplish this. A number of small charges properly located will damage the upper flange, the lower flange, and the web, which will make repair difficult and uneconomical. The twisting of such members in dropping the span and any other feasible method of further destruction should also be considered.

(6) The nature of the terrain under the bridge is of great importance to the success of the demolition. If the distance from the river bed, for example, to the bridge is adequate, the weight of the bridge may be exploited to assist in its destruction (fig. 127).

109. Destruction of Substructures

a. Concrete and Masonry Abutments.

(1) *Charges in fill behind abutment.* The placing of charges in the fill behind an abutment has the advantages of economy in the use of explosives and of concealment of the charges from the enemy until they are detonated. This method also has its disadvantages, as the charges are difficult to place. Where speed is required, charges are not placed behind the abutment if the fill is known to contain large rocks. If the bridge approach is an embankment, the most practical method may be to place explosive charges in a tunnel driven into the side.

(*a*) *Abutments 5 feet or less in thickness and 20 feet or less in height* (fig. 119). Such abutments are demolished by a line of 40-pound cratering charges placed on 5-foot centers in holes 5 feet deep and 5 feet behind the face of the abutment ("triple-nickel-forty"). The first hole is placed 5 feet from one side of the abutment and this spacing is continued until a distance of 5 feet or less is left from the last hole to the other side of the abutment. The formula for computing the number of charges is

$N = \frac{W}{5} - 1$, where N = number of charges and W the width of the abutment. If the wing walls are strong enough to support a rebuilt or temporary bridge, they too should be destroyed by placing charges behind them in a similar fashion.

(*b*) ***Abutments more than 5 feet thick and 20 feet or less in height.*** Such abutments are destroyed by breaching charges placed in contact with the back of the abutment (fig. 120). These charges are calculated by means of the breaching formula, $P = R^3KC$ (para 86*a*), using the abutment thickness as the breaching radius R. The charges are placed at a depth equal to or greater than R. The number of charges and their spacing are determined by the formula $N = \frac{W}{2R}$.

(2) ***Combination charges.*** A combination of external breaching charges and fill charges may be used to destroy abutments more than 20 feet high. Breaching charges placed along the bottom of the abutment face should be fired simultaneously with the charges in the fill behind the abutment. These fill charges may be breaching charges as explained in (1) (*b*) above, or the "triple-nickel-forty" charges depending on the abutment thickness. This tends to overturn and completely destroy the abutment.

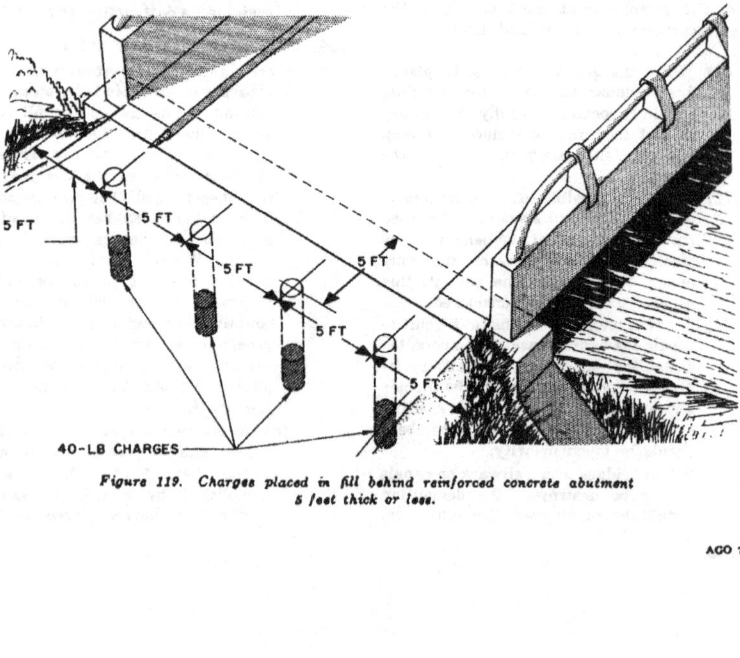

Figure 119. Charges placed in fill behind reinforced concrete abutment 5 feet thick or less.

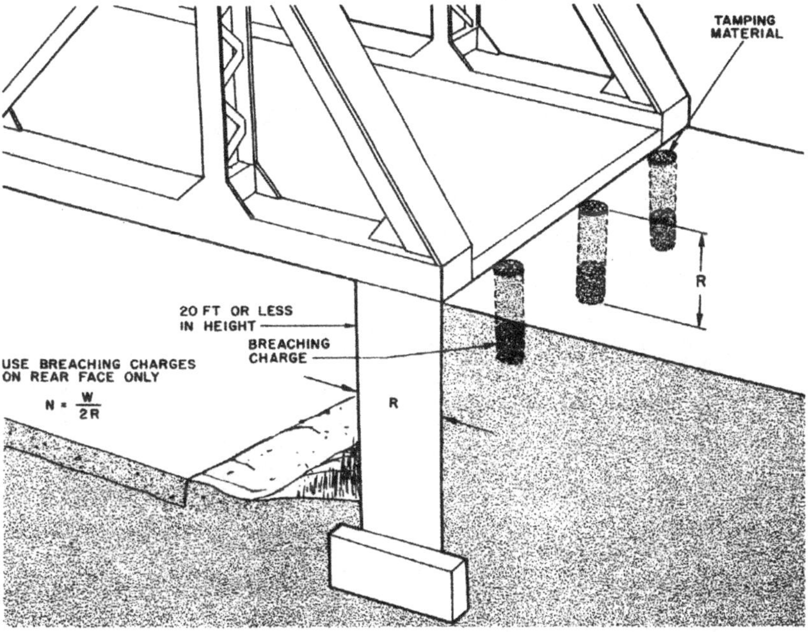

Figure 120. Charges placed in fill behind reinforced concrete abutment more than 5 feet thick.

b. Intermediate Supports.

(1) *Effectiveness.* The destruction of one or more intermediate supports of a multispan bridge is usually the most effective method of demolition (fig. 121). The destruction of one support will collapse the spans on each side of it, so that destruction of only alternate intermediate supports is sufficient to collapse all spans. For repair this will require either the replacement of those supports or the construction of long spans.

(2) *Concrete and masonry piers.* Concrete and masonry piers are demolished either by internal or external charges (fig. 121). Internal charges require less explosive than external charges, but because they require a great amount of equipment and time for preparation, they are seldom used unless explosives are scarce or the pier has built-in demolition chambers. The number of charges required is calculated by the formula $N = \frac{W}{2R}$ (para 87*b*). The size of each charge is calculated by the breaching formula, $P = R^3KC$ (para 86*a*).

(a) *Internal charges.* Plastic (C3 or C4), dynamite, and other explosives are satisfactory for internal charges. All charges of this type should be thoroughly tamped with blunt wooden tamping sticks, *not* with steel bars or tools. If there are no demolition chambers, charges are placed in boreholes, which are blasted by means of shaped charges or drilled with pneumatic or hand tools. A 2-inch diameter borehole holds about 2 pounds of explosive per foot of length or depth. The steel reinforcing bars make drilling in heavily reinforced concrete impractical, however.

(b) *External charges.* External charges may be placed at the base of a pier or higher and spaced not more than twice the breaching radius (para 86*b*) apart. All external charges should be thoroughly tamped with earth and sandbags if time and the size, shape, and location of the target permit.

110. Stringer Bridges

a. Use. The stringer bridge (fig. 122) is the most common type of fixed bridge found in most parts of the world. It is frequently used in conjunction with other types of spans. The stringers are the load-carrying members, while the floor is dead load. Stringers may be timber, concrete, rolled steel sections, or plate girders.

b. Simple Spans. In simple span stringer bridges, the stringers extend only from one support to the next. The method of destruction for this type of superstructure is to place the charges so that they cut the stringer into unequal lengths in order to prevent reuse (fig. 122).

c. Continuous Spans. Continuous spans have continuous lateral supports that extend over one or more intermediate supports. Because the spans are stiffer over piers than at midspan, they may frequently remain in place even though completely cut at midspan. Steel or reinforced concrete is commonly used for such lateral supports. Continuous steel beams, girders, or trusses may be identified readily because they are either the same depth or deeper over piers than elsewhere, and there is no break or weak section over the supports. The superstructure may be demolished by cutting *each member* in *two* places *between supports* and then dropping completely the portion between the cuts. Also, the span may be cut in unequal portions on the sides of the support for overbalancing and falling. Continuous con-

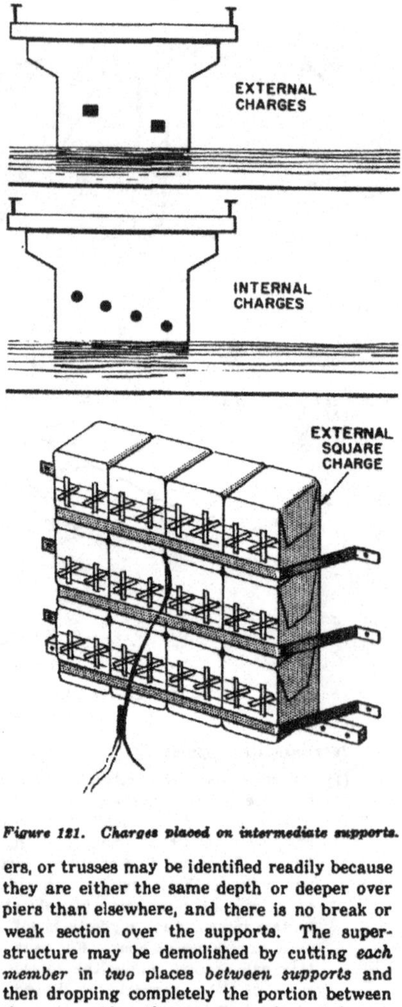

Figure 121. Charges placed on intermediate supports.

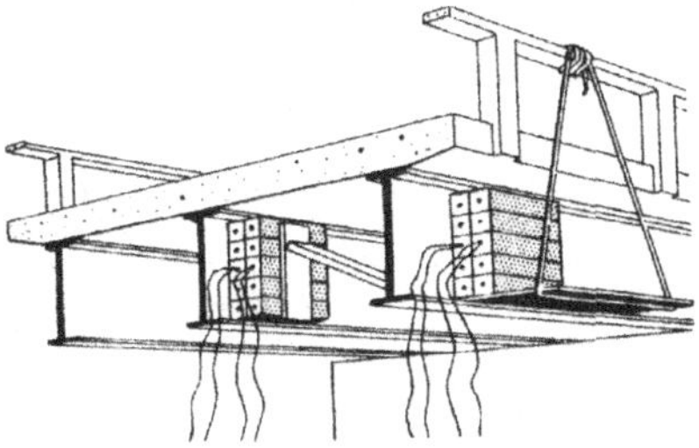

Figure 122. Placement of charges on steel stringer bridge.

crete T-beams or continuous concrete slab bridges may be recognized by the absence of construction or expansion joints over the supports.

111. Slab Bridges

The superstructure of a slab bridge consists of a flat slab support at both ends (fig. 123). This is usually made of reinforced concrete, but may also be of laminated timber or a composite section of timber with a thin concrete wearing surface. If they are simple spans, the superstructure may be destroyed by the use of a single row of charges placed either across the roadway or underneath the span. The breaching formula is used for reinforced concrete slabs; and the timber-cutting (external charge) formula is used for laminated timber. On reinforced concrete slabs, the charges are placed twice the breaching radius apart; and on laminated timber, twice the slab thickness apart. Continuous slab spans must be cut in two places to insure the dropping of the slab or cut in places over the support to provide overturning by unequal weight distribution.

Figure 123. Placement of charges on slab bridge.

112. Concrete T-Beam Bridges

A T-beam bridge is a heavily reinforced concrete stringer bridge with the floor and stringer made in one piece. The floor acts as part of the beam. This type is heavily reinforced. T-beam bridges are generally simple span or continuous span.

a. Charge Placement on Simple Span. Simple span T-beam bridges are destroyed by explosives calculated and placed by the pressure formula or breaching formula.

b. Charge Placement on Continuous Span.

Continuous span T-beam bridges are destroyed by breaching. Charges calculated by the breaching formula are usually placed under the deck in order to use the thickness of the beam *R*. Continuous T-beam bridges may be recognized by the haunching or deepening of the section adjacent to the interior supports. According to conditions, it may be necessary to demolish the piers, demolish the junction between span and pier, or remove all spans by cutting them at approximately one-quarter of their length from each end between supports. Breaching charges are used in all these cases. They may be placed on the roadway, or underneath it if the bridge must be used after the charges have been prepared (fig. 124).

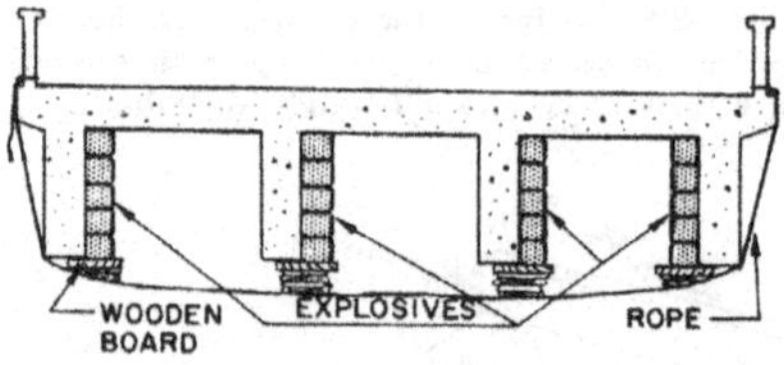

Figure 124. Placement of charges on continuous T-beam span.

113. Concrete Cantilever Bridges

a. Description. Concrete cantilever bridges are identified by the construction joints that appear in the span but not over the piers. Figure 125 shows a cantilever bridge with a suspended span and figure 126, a cantilever bridge without a suspended span.

b. Concrete Cantilever Bridges with Suspended Span. The superstructure of this bridge may be demolished by cutting each cantilever arm adjacent to the suspended span. If a large gap is desired, the cantilever arms should be cut in such a way as to drop the cantilever arms and the suspended spans (fig. 125).

c. Concrete Cantilever Bridges without Suspended Span. As in the bridges above, the superstructure of a cantilever bridge without suspended span is demolished by destroying the cantilever action and unbalancing the cantilever arms (fig. 126). A bridge of this type must be studied to determine the function of the members. Otherwise the charges may not be properly placed.

114. Truss Bridges

a. Description. A truss is a jointed frame structure consisting of straight members (steel or timber) so arranged that the truss is loaded only at the joints. Trusses may be laid below the roadway of the bridge (deck-type trusses) or partly or completely above the roadway (through-type trusses).

b. Single Span Trusses. Single span trusses extend only from pier to pier, usually having a pin joint on one end and a sliding connection at the other. Simple span trusses may be destroyed by any of the following methods:

(1) Cut the upper chord and lower chords at both ends of one truss in each span on the upstream side. This causes the bridge to roll over; thereby twisting the other truss off its support (rotation method). The height of the bridge above the riverbed, however, must permit this. Place the charges on the upper chord so that upon firing the severed upper member will not hang on the lower member and the gap will extend the width of the roadway (fig. 127). If the truss is too small and too light to twist free, both ends of both trusses on each span should be cut or the method described in (2) below should be used.

(2) Cut the upper chords, lower chords, and diagonals of both trusses and the roadway midspan (fig. 128). This is a more complete demolition and makes the reuse of the truss extremely difficult.

(3) Cut both trusses into segments (fig. 129).

c. Continuous Span Trusses. Continuous span trusses are usually extended over two spans, rarely over three. The heaviest chord sections and the greatest depth of truss are located over the intermediate supports. One method of demolition is shown in figure 130. In general, aside from the exact location of charges, the methods given for the destruction of simple span trusses are applicable to con-

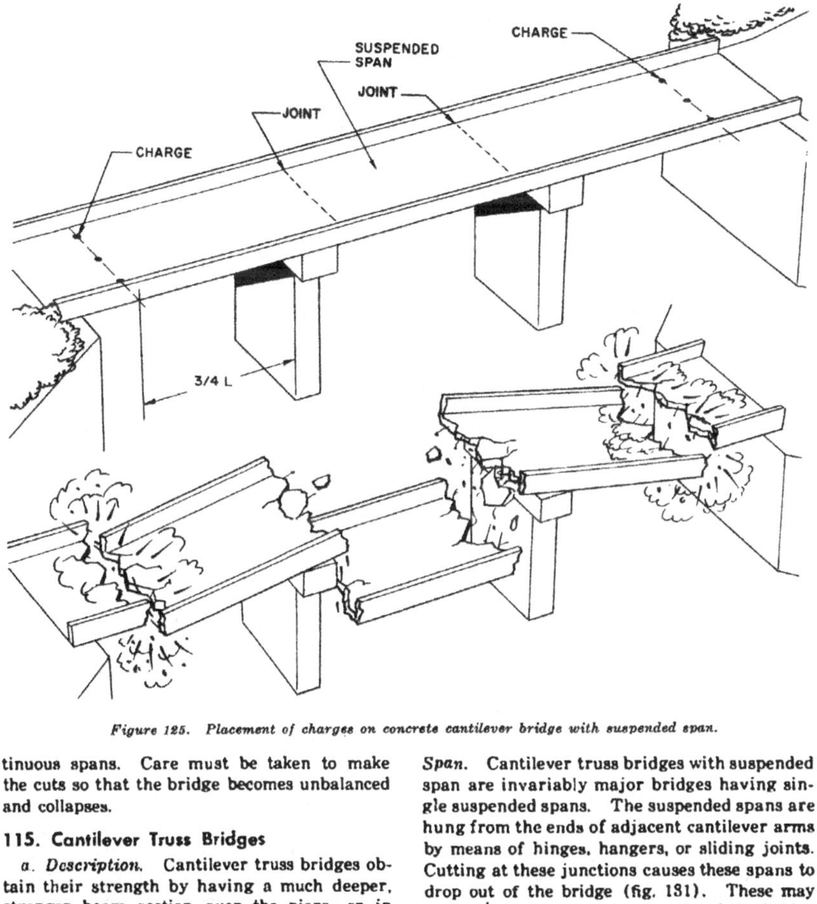

Figure 125. Placement of charges on concrete cantilever bridge with suspended span.

tinuous spans. Care must be taken to make the cuts so that the bridge becomes unbalanced and collapses.

115. Cantilever Truss Bridges

a. Description. Cantilever truss bridges obtain their strength by having a much deeper, stronger beam section over the piers, or in effect, two "arms" that reach partially or completely across the adjacent spans. As cantilever truss bridges are a modification or refinement of continuous truss or continuous beam bridges, the demolition methods given in paragraphs 112 and 114 apply.

b. Cantilever Truss Bridges with Suspended Span. Cantilever truss bridges with suspended span are invariably major bridges having single suspended spans. The suspended spans are hung from the ends of adjacent cantilever arms by means of hinges, hangers, or sliding joints. Cutting at these junctions causes these spans to drop out of the bridge (fig. 131). These may be identified by a thorough study of the bridge structure. Additional steel members may be provided for stabilization but carry no load. The cantilever arms may also be destroyed by the method described in *c* below.

c. Cantilever Truss Bridge Without Suspended Span. To destroy a cantilever truss

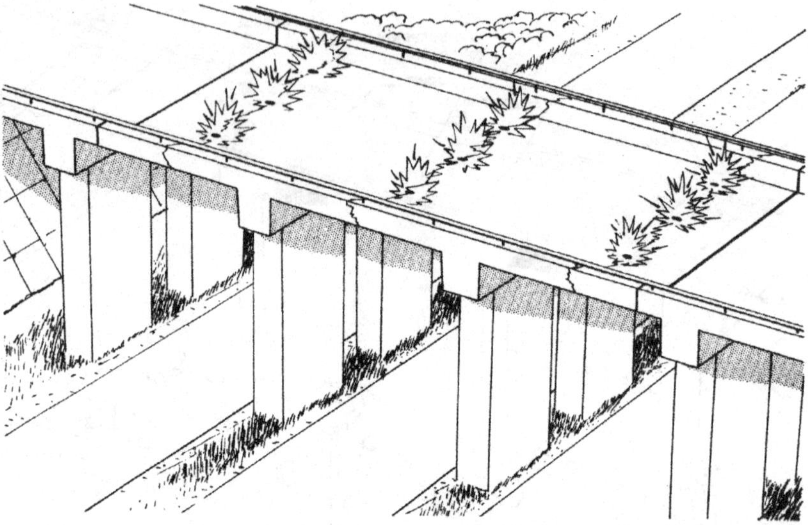

Figure 126. Placement of charge on concerel cantilever bridge without suspended span.

not containing a suspended span, the method shown in figure 132 is recommended. The top and bottom chords are cut at any desired point, and the bridge is cut through near the joint at the end of the arm in the same span. Another method of destruction is to cut completely through the bridge at any two points in the same span, thereby dropping out the length of bridge between the two cuts.

116. Arch Span Bridges

a. Components. A few of the components of bridge arches are described below and illustrated in figure 133.

(1) *Span.* The horizontal distance from one support of an arch to the other measured at the spring line.

(2) *Spring lines.* The points of junction between the arch and the supports.

(3) *Rise.* The vertical distance measured from the horizontal line connecting the supports to the highest point on the arch.

(4) *Crown.* The highest point on the arch.

(5) *Abutments.* The supports of the arch.

(6) *Haunches.* Those portions of the arch that lie between the crown and the spring lines.

(7) *Spandrels.* The triangular-shaped areas between the crown and abutment and above the haunches.

b. Filled Spandrel Arch. A filled spandrel arch consists of a barrel arch (comparatively short span) supporting an earth or rubble fill between the retaining walls. The arch is the most vulnerable at its crown, where it is the thinnest and the earth fill is usually only a foot or two thick. Filled spandrel arches are constructed of masonry (stone or brick), reinforced concrete, or a combination of these

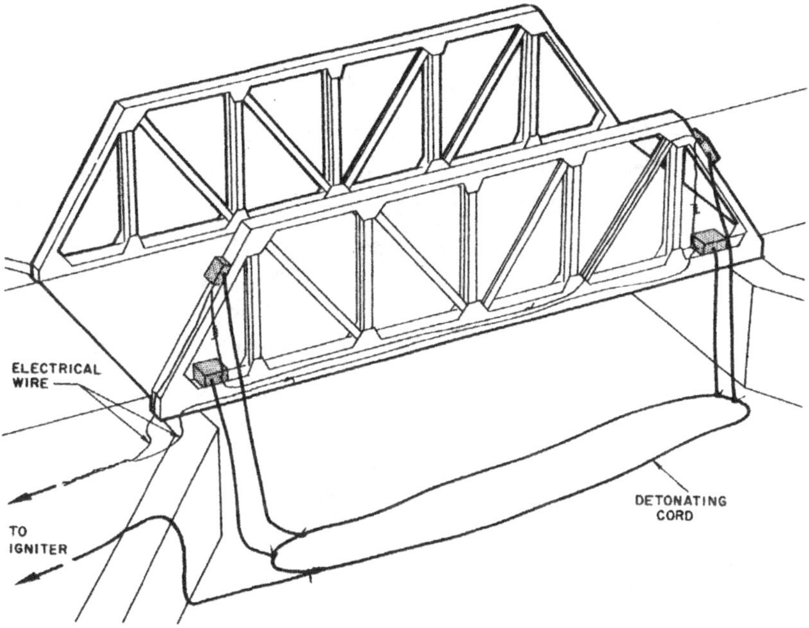

Figure 127. Bridge destruction aided by its own weight.

materials. They may be destroyed by either crown or haunch charges.

c. Demolition by Crown Charges. Crown charges are more easily and quickly placed than haunch charges; but their effectiveness is substantially less, particularly on an arch with a rise that is large in comparison with the span. Crown charges are more effective on the flatter arches because the flatter shape permits a broken portion of the arch to drop out of the bridge. Breaching charges are placed as shown in figure 134.

d. Demolition by Haunch Charges. Breaching charges may be placed at the haunches (just ahead of the abutment) as shown in figure 135 and the traffic maintained until they are fired. If the bridges have demolition vaults or chambers built into the haunches, the charges should be placed there. The presence of demolition vaults is usually revealed by the ventilating brick or steel plate laid in the side wall of the arch. Charges placed in the haunch on the left side will drop out that portion of the arch between lines C and D as shown in figure 135. Charges in both haunches will drop out that portion of the arch between lines C and E. The breaching charges must be placed on the arch ring either in holes on the top or supported on the under side.

117. Open Spandrel Arch Bridges

An open spandrel arch consists of a pair of arch ribs that support columns or bents which in turn support the roadway. The number of arch ribs may vary, and on rare occasions the spandrel bents may be placed on a full barrel

Figure 128. Placement of charge to cut diagonal and upper and lower chords.

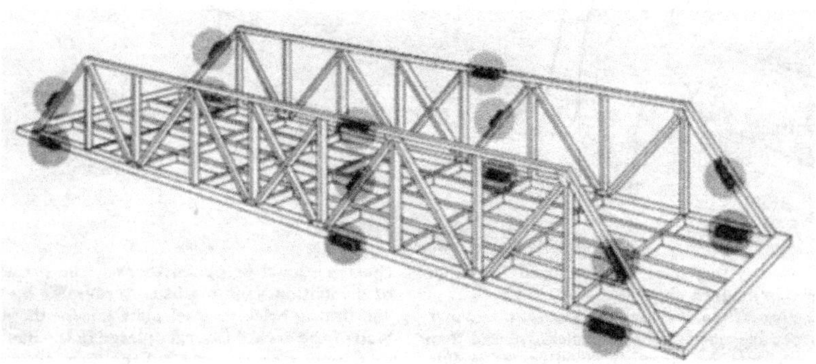

Figure 129. Placement of charges to cut trusses into segments.

arch similar to that which supports the filler material of the filled spandrel arch. The open spandrel arch bridge (fig. 136) may be constructed of reinforced concrete, steel, timber, or any combination of those materials.

a. Demolition of Concrete Open Spandrel Arch. The ribs of a concrete open spandrel arch bridge (fig. 136) are about 5 feet wide. The thickness of the arch rib at the crown varies from about 1 foot for spans of 50 to 60 feet in length to 3 feet for spans of 200 feet or more. The arch thickness at the spring line is ordinarily about twice the thickness at the crown. In long spans, the ribs may be hollow. The floor slab is usually close to the crown, permitting packing of charges against the rib at this point. Here again, the same difficulties are found in reaching the working points at the crown as in T-beam (para 112) or in stringer bridges (para 110). Since for struc-

Figure 130. Demolition of a continuous span truss.

Figure 131. Demolition of cantilever truss with suspended span.

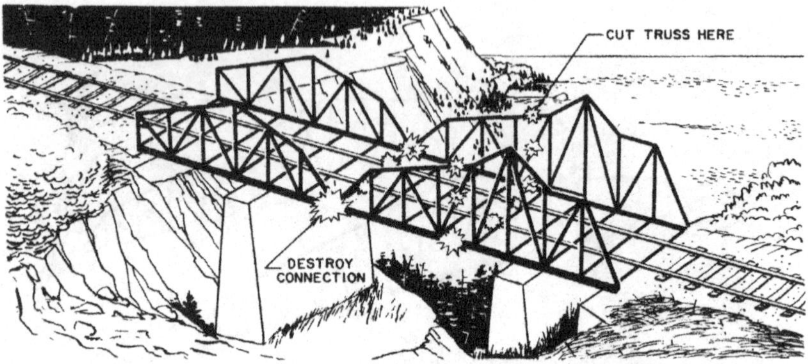

Figure 132. Demolition of cantilever truss without suspended span.

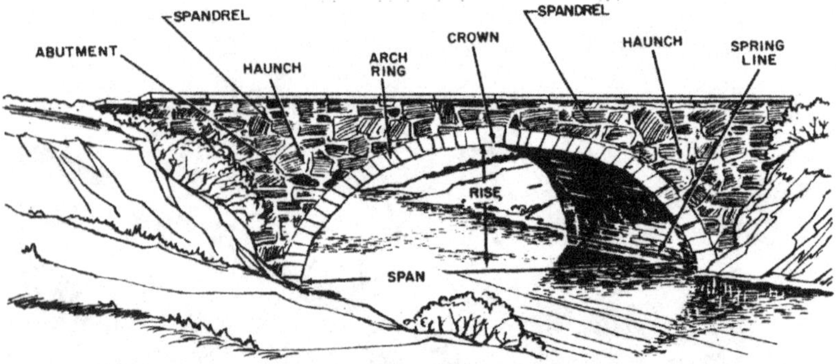

Figure 133. Arch components.

tural reasons, the haunch over the abutment is most likely to be heavy, effective destruction of the arch itself by means of light crown charges may leave a substantial pier at roadway level in an undamaged condition. This type of structure is usually built in one massive unit rather than in lighter separate component parts and is very tough. Also, cutting the span at each end may drop the whole span only a relatively short distance. This may make the damaged bridge an excellent support for building a new temporary bridge. Therefore to prevent utilization of such a span, one charge is placed at the haunch and another at the crown. The uncut half-span will then also fall if the total span exceeds 50 or 60 feet. The charge at the haunch is computed for placement at either the ring or the pillar over

Figure 134. Breaching by crown charges on filled spandrel arch bridge.

the support, whichever has the greater radius. For short single arch spans, destroy the entire span with breaching charges laid behind the abutments or behind the haunches.

b. Demolition of Steel Arch Span. Steel arches are of four general types: continuous arches ((1) fig. 137), one-hinged arches (2), two-hinged arches (3), and three-hinged arches (4). One-hinged arches are hinged in the middle; two-hinged arches, at both ends; and three-hinged arches, at both ends and in the middle. Continuous arches and one-hinged arches are destroyed by placing charges at both ends of the span just far enough from the abutment to allow the arch to fall. Two-hinged and three-hinged arches need only one charge apiece for destruction. This should be placed at the center of the span.

118. Suspension Span Bridges

The suspension span bridge is usually a major bridge distinguished by two characteristics: the roadway is carried by a flexible member, usually a wire cable, and the spans are long (fig. 138).

a. Components.

(1) *Cables.* Cables of suspension bridges are usually two steel multi-wire members that pass over the tops of towers to anchorages on each bank. The cables are the load-carrying members. (The "Golden Gate" bridge has 127,000 miles of cable wire.)

(2) *Towers.* Towers of a suspension bridge support the cables or load-carrying members. They may be made of steel, concrete, masonry, or a combination of these materials.

(3) *Trusses or girders.* The trusses or girders of a suspension bridge do not support the load directly. They provide stiffening only.

Figure 135. Breaching by haunch charges on filled spandrel arch bridge.

Figure 136. Demolition of reinforced concrete open spandrel arch bridge.

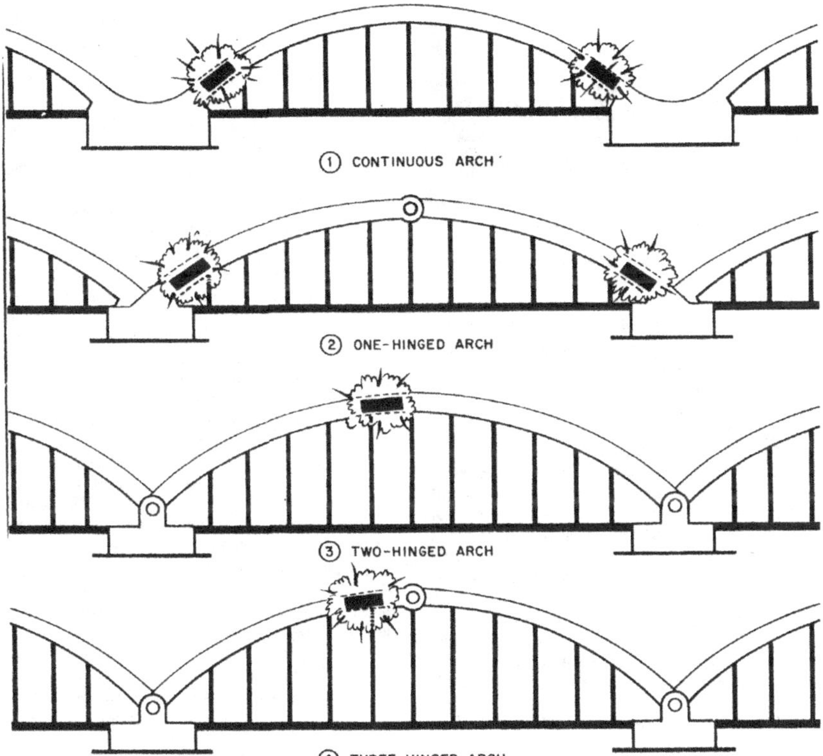

Figure 137. Demolition of steel arch bridges.

(4) *Anchorage.* The usual anchorage is merely the setting of the splayed end of the cable in a rock or a concrete mass. This may be large-sometimes as much as 1000 cubic feet in size.

b. Destruction.

(1) *Major structures.* The towers and anchorages of a major suspension bridge are usually too massive to be destroyed, and the cables are too thick for positive cutting with explosives. The most economical method of destruction is either by dropping the span leading onto the bridge or dropping a section of the roadway by cutting the suspenders of the main or load-bearing cables. The length of this section should be determined by an analysis of what capabilities the enemy has for repair in the time

Figure 138. Suspension span bridge.

he is expected to retain the site, particularly the erection of a prefabricated bridge. It may also be feasible where there are reinforced concrete towers to breach off the concrete and cut the steel.

(2) *Minor structures.* The two vulnerable points of a minor suspension bridge are the tower and the cables.

(3) *Towers.* Charges may be placed on the towers slightly above the level of the roadway. A section should be cut out of each part of each tower. A charge is placed on each post to force the ends of the cutout section to move in opposite directions twisting the tower. This will prevent the ends of a single cut from remaining in contact. Demolition chambers, provided in some of the newer bridges, make blasting easier, quicker, and more effective.

(4) *Cables.* Charges should be placed on the cables as close as possible to firm support such as at the top of the tower or at an anchorage. Extreme care should be taken to extend the charges not more than one-half the distance around the circumference of the cable. These charges are bulky, exposed, and difficult to place; and the cables are difficult to cut because of the air space between the individual wires. Shaped charges, however, with their directed force effect, may be used to advantage.

119. Floating Bridges

Floating bridges consist of a continuous roadway of metal or wood supported by floats or pontons.

a. Pneumatic Floats. Pneumatic floats consist of rubberized fabric made into airtight compartments and inflated with air.

(1) *Hasty method of destruction.* The anchor cables and bridle lines may be cut with axes and the steel cable, by explosives.

(2) *Deliberate method of destruction.* The floats may be punctured by small arms or machinegun fire. This requires a considerable volume of fire because of the large number of watertight compartments in each float. Detonating cord stretched snugly across the surface of inflated ponton compartments will make a clean cut through the material. One strand will suffice to cut most fabrics; two may be required for heavier material. Also one turn of detonating cord around an inflation valve cuts it off at the neck or does other damage. Lines placed around valves should not be main lines but branch lines run off from the main line, as the blast wave may fail to continue past the sharp turn.

b. Rigid Pontons. Rigid pontons are made of various materials such as wood, plastic, or metal. Most of these are open but occasionally they are decked over.

(1) *Hasty method of destruction.* A ½-pound charge of explosive is placed on the upstream end of the bottom of each ponton and detonated simultaneously. If the current is rapid, another method is to cut the anchor cables so that the bridge will be carried downstream.

(2) *Deliberate method of destruction.* The bridge is severed into rafts and half-pound charges of explosives are placed at each end of each ponton and detonated simultaneously.

c. Treadways. Charges to destroy the treadway of any metal treadway type of floating bridge may be calculated by means of the steel-cutting formula. The placement and amount of the charges to be used depends on the type of bridge to be destroyed. In general, if charges are set to sever the roadway completely at every other joint in the treadway, the bridge will be damaged beyond use.

120. Bailey Bridges

A 1-pound charge placed between the channels of the upper and lower chords will destroy the panels. A ½-pound charge will cut the diagonals and a 1-pound charge, the sway bracing (fig. 139).

a. Bridge in Place.

(1) The bridge is severed into parts by cutting panels on each side, including the sway braces. The line of cut is staggered through the panels; otherwise the top chords may jam and prevent the bridge from dropping. In double-story or triple-story bridges, the charges are increased on the chords at the story junction line.

(2) For further destruction, charges are placed on the transoms and the stringers.

b. Bridges in Storage or Stockpile. Destruction of bridges in storage must be such that the enemy cannot use any of them as a unit or any parts for normal or improvised construction. This requires that one essential component, not easily replaced or improvised, be destroyed so that the bridges at a particular stockpile cannot be used. In this way it will also be impossible for the enemy to obtain replacements for other sectors. The component that fulfills all of these conditions is the panel. To make the panel useless, the female lug in the lower tension cord is removed or distorted. All panels should be destroyed before other components are destroyed.

Section IV. DAMAGING TRANSPORTATION LINES

121. Highways

Disruption of enemy transportation lines is an important demolition objective. The extent of demolition, however, depends upon the analysis of the system and the mission. By the destruction of the road net, the attacking forces are halted or delayed, the movement of supplies is prevented, and frequently new construction is required before the enemy can advance. This may be accomplished by the demolition of bridges, by blowing road craters, by placing wrecked items of equipment and debris in cuts and defiles, and by the construction of abatis and roadblocks.

122. Railroads

a. Tracks.

(1) If possible, the destruction of railroads with explosives should be done at vulnerable points. These are curves, switches, frogs, and crossovers, which may be destroyed with a small amount of explosive. This is called the "spot" method. Placement of charges is shown in figure 140.

(2) A length of single track may be destroyed rapidly by a detail of soldiers with a push car, ¼-ton truck, or other vehicle supplied with explosives, nonelectric blasting caps, time fuse, fuse lighters, and filled sandbags. Several soldiers ride the vehicle, prime 1-pound charges, and hand them, together with sandbags, to men walking immediately behind the vehicle. These men place the charges against the rail on alternate connections of both tracks for a distance of about 500 feet, and then tamp them well with sandbags. Tamping is not required to break the rail, but will destroy a longer length of rail. Other men follow about 250 yards behind the vehicle to light the fuses. This method requires approximately 20 pounds of explosive per 500-foot length of single track line. It should be repeated at approximately 1½-mile intervals. Such procedure is particularly advantageous when the supply of explosives is limited or when time or other factors prohibit complete destruction of a line. It causes a greater delay in repair than a concentrated destruction of short lengths of track. If time, explosives and other conditions permit, however, complete

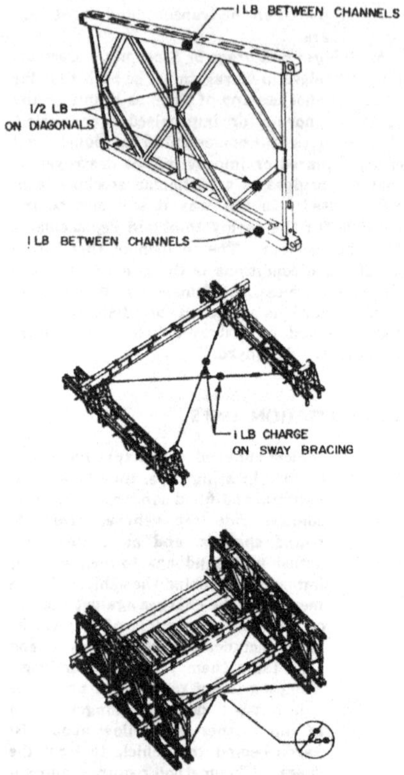

Figure 139. Demolition of Bailey bridge.

destruction of long lengths of track will provide maximum delay.

(3) Tracks may be made unserviceable without the use of explosives by tearing up sections of the track, especially along cuts, fills, or embankments, where the use of reconstruction equipment is restricted and work areas are limited. This may be done by removing fishplates from both ends of a section of track, fastening a neavy chain or cable to it, and pulling it up by a locomotive. Also, a large hook towed by a locomotive is useful to tear up ties and rails. Whenever possible, ties loosened from the rails should be piled and burned.

b. Roadbeds. Roadbeds are damaged by the methods used in making road craters and antitank ditches.

123. Tunnels

Railway and highway tunnels located on major routes to strategic industrial or military areas are vulnerable to demolition and therefore desirable targets. Tunnel demolition, however, with hastily placed conventional explosives is impossible unless huge quantities are used. But when demolition chambers exist or time, men, and equipment are available, considerable damage to tunnels can be accomplished with reasonable amounts of explosive.

a. Principal Factors in Tunnel Demolitions. The most critical factor in tunnel demolition is the tightness of the lining against solid rock. The actual thickness and strength of the lining are of secondary importance. The degree of contact of the walls with surrounding rock influences the amount of blast energy transmitted to the rock or retained in the concrete and the consequent movement of broken fragments, which may permit their being dislodged and dropped into the tunnel.

b. Hasty Demolitions. The hasty demolition of tunnels with reasonable amounts of conventional high explosives is ineffective. No hasty method has yet been devised that will cause extensive damage. The enemy may be temporarily deprived of the use of a tunnel by breaking and dislodging portions of the lining with normal breaching charges placed at a number of points and by creating slides at the tunnel portals by placing cratering charges in the slope above them. Nuclear devices of proper size advantageously placed, will effectively demolish a tunnel.

c. Deliberate Demolitions. Deliberate tunnel demolitions will produce satisfactory results when explosive charges are detonated in prepared chambers in the material adjacent to the inner face of the tunnel, whether it is lined or

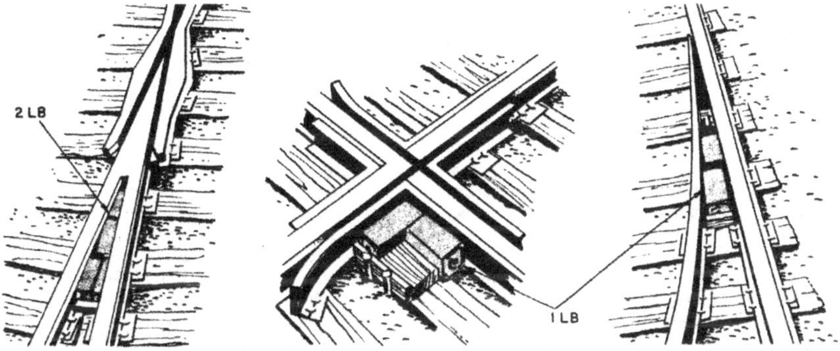

Figure 140. Demolition of railroad switches, frogs, and Crossovers.

not. Excessive charges are not required. Maximum damage desired in any tunnel is that of obstructing it with broken rock. Secondary damage by fire may also occur. Caving, which may result from structural damage to the tunnel arch, cannot be predicted. To obtain maximum damage to tunnels, the methods outlined below are adequate.

(1) *Charge chambers.* Tunnel charge chambers should be so constructed that each chamber parallels the long axis of the tunnel at or above the spring line. The T-design tunnel charge chamber is an efficient means of inflicting serious damage. The chambers may be constructed opposite each other at staggered intervals on opposite sides, or all on one side of the tunnel. The maximum burden, which is the distance from the charge to the inside rock wall, should be 15 feet. The tunnel charge chamber should be no larger than necessary for convenience of construction and loading and no smaller than 3 feet wide by 4½ feet high. Charge chambers should be constructed far enough inside the tunnel portal to insure confinement of the charge. The minimum of side hill or outside burden should be 30 feet.

(2) *Charges.* Seven hundred and fifty pounds of high explosive is an effective minimum charge for single placement within a T-type chamber of 15-foot burden. Charges should be placed on 30-foot centers.

(3) *Stemming.* Stemming, the material with which a borehole or charge chamber is filled or tamped (usually earth-filled sandbags), is necessary. It should extend from the last charge in the T-type chamber to the chamber entrance. Stemming is not necessary, however, between charges within the chamber.

d. Deliberate Demolition of Tunnels With Prepared Charge Chambers. Some tunnels have chambers or holes in the roof for the purpose of demolishing them at some future time. Their presence is usually indicated by the open brick ventilators placed over them. If no ventilators are present, these chambers may be located by striking the roof of the tunnel with some heavy metallic object, which usually produces a hollow sound. Explosives, compacted as tightly as possible, are placed in the chambers, which are then closed and sealed except for the place where the fuze or firing wire passes through. Sandbag stemming is recommended in all charge chambers in timbered tunnels, as the sandbags increase the possibility that the timbers may ignite when the charge is detonated.

124. Water Transportation Systems

The extent of demolition depends largely upon the mission, materials available, and an analysis of the system as to how critical, accessible, repairable, and vulnerable it may be.

a. Vessels, Piers, and Warehouses. Vessels can seldom be destroyed efficiently by landbased troops, unless they are tied up at docks, piers, and warehouses, however, are excellent demolition targets, especially for the use of fire.

b. Channels. The most expeditious way to block a channel is to sink a ship or a loaded barge at a point that cannot be bypassed. Channels with retaining walls may be blocked effectively by detonating breaching charges behind the retaining walls.

c. Dams. Since a prohibitive amount of explosive is generally required to destroy an entire dam structure, the best and quickest method is to destroy the machinery and the equipment. If the purpose of the demolition is to release the water in the dam, all that is required is to destroy the gates on the crest of the dam, the penstocks or tunnels used to bypass the dam or to carry water to hydroelectric plants, or the valves or gates used to control the flow in the penstocks or tunnels. In dams, partly or wholly constructed of earth fill, it may be possible to ditch or crater down below the existing waterline and thus allow the water itself to further erode and destroy the dam. Nuclear devices may also be used to advantage.

d. Canals. In most cases, a canal may be made useless by destroying the lock gates and the operating mechanism that controls them. This mechanism, which includes the electrical equipment and perhaps pumps, is the easier to destroy and should therefore be attacked first. If time permits, the gates themselves may be destroyed. The lock walls and canal walls may be destroyed by breaching charges or cratering charges placed behind them.

125. Airfields

Airfields may be destroyed by ADM. They may be rendered unusable by cratering runways or placing objects on surfaces to prevent use by aircraft. The destruction of POL and munitions stocks and of repair and communications facilities is also effective. Rooters, plows, and bulldozers can ditch runways that are not constructed of concrete. Friendly operational and nonoperational airfields should be destroyed only in areas where the resulting wreckage will provide the maximum impediment to enemy movement and operations. They should, however, be made ready for demolition only during the preparation for an organized withdrawal when seizure by the enemy is imminent.

a. Plans for Demolition.

(1) The methods of destroying any airfield depend on the materials at hand, the type of installation to be destroyed, and the time and equipment available to complete the job. Aircraft and equipment may be destroyed instantly by directing weapons fire against them. Whenever possible, bombs and ammunition should be used as explosive charges (app C). Gasoline and other fuels (POL products) may be used to aid in the destruction, by fire, of vehicles, equipment, and buildings.

(2) When time permits, a detailed plan for demolition of the airfield should be prepared before any charges are placed. This should include—

(*a*) Location of charges.

(*b*) Type of explosives.

(*c*) Size of each charge.

(*d*) Priority in preparation and placement of each charge.

(*e*) Total amount of itemized explosives and other materials needed to effect demolitions included in the plan.

(*f*) Assignment of personnel or groups to prepare specific charges.

b. Priorities for Demolition. It is seldom possible to destroy an airfield completely because of the great amount of explosives and time required. Thus it is necessary to determine what specific demolition is to be done and in what order specific operations are to be accomplished. Airfield demolition plans should be very flexible, particularly in regard to priorities. The order of priority should vary according to the tactical situation. The following list suggests an order of priority for airfield demolition, which may be modified to suit the tactical situation.

(1) Runways and taxiways and other landing areas.

(2) Routes of communication.

(3) Construction equipment.

(4) Technical buildings.

(5) Supplies of gasoline, oil, and bombs.

(6) Motor vehicles and unserviceable aircraft.

(7) Housing facilities.

c. Runways and Taxiways. Runways and taxiways have first priority in a demolition plan because the destruction of landing surfaces is the most important single item. Whenever possible, demolition of an airfield should be considered during construction by the placing of large conduits in all fills. This requires little extra work and provides a means of placing explosives under the runway. Standard deliberate and hasty craters may be useful in the demolition of runways and taxiways. Shaped charges may be used to breach thick concrete pavement when speed is essential. The placing of individual cratering charges diagonally down the runways or taxiways, or in a zigzag line running diagonally back and forth, provides more complete destruction. When pierced steel plank or other type of landing mat is used on an airfield, substantial damage may be done by attaching a large hook to sections of the mat and pulling it out of place with a tractor. This should be followed by cratering. A hasty, satisfactory obstacle may be produced by the use of 40-pound cratering charges spaced every 15 feet across the runway and buried 4 feet under the ground.

d. Turf Surfaces and Pavements. Bituminous surface treatments or thin concrete pavements can be destroyed by bulldozers, graders, and rooters. Turf airstrips can be destroyed by plowing or cratering.

e. Aircraft. Unserviceable aircraft should be destroyed by the detonation of 4 pounds of TNT placed on each crankshaft between the propeller and the engine and 1 pound of TNT placed on the instrument panel to prevent salvage. The engines of jet-propelled aircraft should be destroyed by detonating charges on essential parts, such as the compressor, air intake, or the exhaust turbine. Radio equipment, bombsights, radar, and tires should be removed or destroyed.

126. Pipelines

The most vulnerable points of a pipeline system are the storage tanks and pumping stations.

a. Storage Tanks. Storage tanks filled with fuel may be destroyed most effectively by burning with incendiary grenades or the burst of .50-caliber incendiary ammunition. Empty tanks may be destroyed by detonating charges against the base.

b. Pumping Stations. Booster pumping stations on cross-country pipelines, being very vulnerable, should be destroyed. Gravel or other solid objects introduced into the pipeline while the pumps are running will damage the moving parts, although not to the degree possible with explosives. If time permits, the pumping station should be burned after the equipment has been destroyed by explosives.

c. Pipe. The pipe used in pipelines is destroyed only during scorched earth operations because of the great amount of effort necessary for effective damage. Junctions, valves, and bends are the most suitable points, particularly when the line is buried. Another method is to close all valves on the line; the expansion that occurs, even in subzero weather, will break it.

Section V. DAMAGING COMMUNICATION SYSTEMS

127. Telephone and Telegraph Lines

Although damage to an enemy telephone system or telegraph system may never be extensive, it does have a great delaying effect. Telephone and telegraph switchboards and instruments are the best points of attack. Generally 1-pound charges placed on the cables are adequate to sever them. Also dial systems may be damaged by smoke from burning oil. Pole lines are not satisfactory targets as they are strung over long distances and can be destroyed only in spots. They may be made temporarily useless by cutting or grounding the wires or by cutting the poles with small external timber-cutting charges and then burning. The wire should be cut into short lengths to prevent further use.

128. Radio Installations

Radio provides rapid communication between far distant points that would otherwise be without communication. Antenna towers, usually constructed of steel and braced with guy wires, are the most accessible part of any radio installation. They are destroyed by cutting the guy wires and by placing cutting charges against the base. The towers should be toppled over the transmitter station or across the high voltage transmission line through which the installation received its power. Equipment and standby power units may be destroyed by mechanical means or by demolition charges. Transformers also are very vulnerable. They will burn themselves up if a hole is blasted in the side or bottom and the oil let out, provided they do not have automatic thermo cutoff switches.

Section VI. DESTRUCTION OF BUILDINGS AND INSTALLATIONS

129. Buildings

Buildings may be destroyed by explosive or other methods. The methods used and the extent of demolition usually depend on the time available.

a. Masonry or Concrete Buildings. Masonry or concrete buildings may be destroyed by breaching charges placed on the inside and at the base of the exterior walls.

b. Wood or Thin-Walled Buildings. Wooden frame buildings may readily be destroyed by fire. Another method is to close all doors and windows and explode on the ground floor a concentrated charge (dust initiator) equal to ¼ pound to 1 pound of explosive per cubic yard of volume (app F). Such buildings may be dismantled, however, if time permits.

c. Steel Framed Buildings. Stink bombs and other malodorous devices and contaminants may be all that is required to damage the interior furnishings or equipment. Another method—to expose the interior to extreme heat, 1000° F for 10 minutes—will cause failure of the structural steel members. Buildings with steel frames may also be destroyed by first breaching the concrete or masonry where necessary to expose the vital steel members and then cutting them with explosive charges.

d. Concrete Beam, Curtain Wall Buildings. Concrete beam curtain wall buildings, constructed in such a way that the load is carried by reinforced concrete beams and columns, are destroyed by placing breaching charges inside the buildings at the base of the exterior wall and at the base of all intermediate columns on the ground floor.

130. Electric Power Plants

Before destruction, electric power plants should be studied so that the amount of damage will be adequate but not exceed tactical demands. They may be destroyed by cutting the windings of generators and motors, by placing and detonating a 2-pound charge inside the casings, or by pouring gasoline on the generators and lighting them. Generators also may be "shorted" out by the use of metal powder or shavings. The shafts of motors and generators are broken. Damage can also be done by removing or contaminating the lubricating oil with metal filings or aluminum powder and then running the machinery. Boilers are burst with a cutting charge. All buildings, transmission towers, penstocks, and turbines of hydroelectric plants may be destroyed.

131. Water Supply

The pumping station, filtration plant, and reservoirs of a water supply system are usually the points most accessible to attack. Storage tanks are demolished by charges calculated on the basis of 1 pound of explosive per 100-cubic foot capacity. The charge is detonated inside the tank when it is full of water. The water acts as a tamping material. Shaped charges are also useful in this capacity. The standoff, however, should be cut down considerably. Wells sunk in soft soils are damaged beyond repair by charges that cut the lining. Wells in rock and hard soils, having little or no lining, are demolished by exploding large breaching charges 6 to 12 feet from the edge of the well and deep enough to secure good tamping. If time does not permit such preparation, a large charge is exploded halfway down against the side.

132. Petroleum, Oil, and Lubricating Refineries

POL refineries are readily demolished, having such extremely vulnerable points as cracking towers, steam plants, cooling towers, and POL stock. These are easily damaged by explosive charges and fire. The demolition of such installations should be planned and executed only by persons familiar with their design and construction or after extensive investigation.

Section VII. DESTRUCTION OF EQUIPMENT AND SUPPLIES

133. Introduction

a. Authority for Destruction. The destruction of friendly materials is a command decision, implemented only on authority of the division or higher commander. Equipment and supplies that cannot be evacuated and may, therefore, be captured by the enemy are destroyed or made unserviceable, except for medical materials and stores, which are not to be intentionally destroyed (DA Pam 27-1 and FM 8-10).

b. Destruction Areas. Whenever possible, mobile equipment is demolished in places where it most effectively impedes the advance of the enemy. Examples of such places are—

(1) Approaches to bridges (fills).

(2) Airfield landing strips.

(3) Cuts, fills, or hills on roads.

(4) Sharp bends of roads.

(5) Roads leading through densely wooded areas.

(6) Narrow streets in thickly populated or built-up areas.

c. Priority of Operations. Destruction must be as complete as the available time, equipment, and personnel will permit. If all parts of the equipment cannot be completely destroyed, the most important ones should be damaged. Special attention must be given to those parts that are not easy to duplicate or rebuild. Particular care must be taken that the same components are destroyed on each piece of equipment; otherwise the enemy may assemble a complete unit with parts taken from several partly destroyed units (cannibalization).

d. Precautions. When material is destroyed by explosives or by weapons fire, flying fragments and ricocheting bullets create a hazard. Thus demolition must be accomplished in an area free of friendly troop concentrations.

134. Planning

Standing operating procedures for all units should contain a plan for the destruction of all equipment and supplies except medical supplies, which are left intact for enemy capture. Such a plan will insure that the maximum and most effective damage is done to materiel and will deny the use of friendly equipment to the enemy. It should outline the required extent of demolition and include priorities of demolition and methods of destruction for all items issued to the unit. If explosives are to be used, the amounts required should be indicated. The plan must be flexible enough in its designation of time, equipment, and personnel to meet any situation. In order to make cannibalization by the enemy impossible, each equipment operator must be familiar with the priority sequence in which essential parts, including extra repair parts, are to be destroyed. He must also be familiar with the sequence to be followed for total destruction

135. Methods of Destroying Material

The following methods of destroying material may be used either singly or in combination. The actual method or methods used in a given situation depend on the time, personnel, and means available.

a. Explosives. All military explosives are effective in destroying equipment.

b. Mechanical Means. Material may be destroyed by mechanical means. Sledge hammers, crowbars, picks, axes, and any other available heavy tools are used to smash or damage whatever is to be destroyed.

c. Weapons Fire. Hand grenades, rifle grenades, antitank rockets, machinegun fire, and rifle fire are a valuable means of destroying materiel.

d. Thermite Grenades. Flammable material and equipment may be destroyed or made unserviceable by heat generated by the thermite grenade. The material should be soaked with fuel before burning.

e. Fire. Rags, clothing, or canvas should be packed under and around the materiel to be destroyed. It should then be soaked with gasoline, oil, or diesel fuel. Damage from fire may not always be as severe as expected. Engine or transmission parts heated to less than a dull red heat are not seriously damaged provided they are lubricated immediately after the fire to prevent corrosion. Electrical equipment, including motor or generator armature windings and other wiring, is effectively destroyed by burning. All parts made from low-melting-point metal may be almost completely destroyed by fire.

f. Water. The damage resulting from submerging equipment in water is not generally very severe, but the method is sometimes rather quickly and easily accomplished. Total submersion also provides concealment of equipment (*h* below).

g. Abuse. Much damage can be done to equipment, particularly to engines, by deliberate improper operation. Such abusive treatment may proceed even after abandonment, if hasty action becomes necessary, by leaving the equipment in an improper operating condition.

h. Concealment. Easily accessible vital component parts of equipment may be removed and scattered through dense foilage, thus preventing or at least delaying use by the enemy. Vital parts of entire items may be hidden by throwing them into a lake, stream, or other body of water (*f* above).

i. Boobytrapping. Boobytraps are placed in debris after destruction is completed, if time permits. See FM 5–31 for the techniques.

136. Destruction of Combat Equipment

There are various publications on the proper methods of destroying military combat equipment. The FM 23-series is concerned with the destruction of small arms such as rifles, pistols, mortars, and ammunition; and the FM 17-series, with the destruction of armored vehicles and their weapons.

137. Training

Training does not involve the actual demolition of any materiel but the simulated breaking of vital parts, the placing of dummy charges, and the selection of sites suitable for the destruction of equipment in order to block communication routes. Drivers and operators should be made familiar with each step in the appropriate method for the destruction of their equipment and supplies. It should be emphasized that in planning destruction operations, the following methods should be considered in the order given:

a. Mechanical damage to vital parts.

b. Use of explosives.

c. Use of weapons fire, fire, and water.

CHAPTER 5

SAFE HANDLING, TRANSPORTATION, AND STORAGE OF EXPLOSIVES

Section I. GENERAL SAFETY PRECAUTIONS

138. Safety Rules and Responsibility

a. Compliance. Safety rules regarding explosives, caps, and demolition equipment will be followed strictly during training. In combat, however, some must of necessity be altered so that unit missions may be accomplished. In all other situations they will be observed to the fullest extent permitted by time, materials available, and requirements of the mission. Also, post regulations and local and units SOP's will be observed.

b. Responsibility. The responsibility of the preparation, placement, or firing of charges must not be divided. One individual should be responsible for the supervision of all phases of the demolition mission.

139. Safe Distance Formula

Distances at which persons in the open are safe from missile hazards from bare charges placed on or in the ground are given in table XIV. The formula for computing safe distances from explosives so placed is—Safe distance in feet $= 300 \sqrt[3]{\text{pounds of explosives}}$. (For quarrying the safe distance in feet $= 350 \sqrt[3]{\text{pounds of explosives}}$). The minimum safe distance for soldiers in a missile-proof shelter is 300 feet (AR 385–63).

Table XIV. Minimum safe distances from explosives for persons in the open

Pounds of explosive	Safe distance in feet	Pounds of explosive	Safe distance in feet
1–27	900	150	1,593
28	909	175	1,678
29	921	200	1,754
30	930	225	1,825
32	951	250	1,890
34	969	275	1,951
36	990	300	2,008
38	1,008	325	2,063
40	1,020	350	2,114
42	1,041	375	2,163
44	1,050	400	2,210
46	1,074	425	2,256
48	1,080	450	2,299
50	1,104	475	2,341
55	1,141	500	2,381
60	1,170	525	2,420
65	1,200	550	2,458
70	1,230	575	2,495
75	1,260	600	2,530
80	1,290	625	2,565
85	1,317	650	2,599
90	1,344	675	2,632
95	1,368	700	2,664
100	1,392	725	2,695
125	1,500	750	2,725

Table XIV.—Continued

Pounds of explosive	Safe distance in feet
775	2,756
800	2,785
825	2,814
850	2,842
875	2,869
900	2,896
925	2,923
950	2,949
975	2,975
1000	3,000

Note. Chart is based on formula:

d $300 \sqrt[3]{P}$.
d safe distance in feet
p pounds of explosive

140. Package Care and Repair

Carelessness, rough handling, and disregard for safety rules cause premature explosions, misfires, and in many cases serious accidents. Issued explosives and auxiliary items are packed in moisture-resistant containers and proper packing boxes to withstand field conditions of transportation and storage. Containers and boxes must never be handled roughly; they must never be broken, cracked, or dented. Some special items, if distorted, lose part of their effectiveness. Damaged packing boxes and containers must be repaired immediately; all defaced parts of marking must be transferred to new parts of the boxes. Broken airtight containers, such as those containing chemical mines, should be destroyed.

Section II. TRANSPORTATION, STORAGE, AND DISPOSAL

141. Transportation

a. Safety Policy. Local transportation of explosives for immediate use is directed by AR 385–63. The Department of the Army clearly defines the safety responsibilities of transportation officers at their installations. Local safety SOP's are provided to insure that all persons participating in the transportation of explosives have proper instruction in safety requirements and are held to account for all violations of procedure.

b. General Rules. The following rules are observed:

(1) Vehicles used for the transportation of explosives shall not be loaded beyond rated capacity and the explosives shall be secured to prevent shifting of load or dislodgement from the vehicle in transit. In all open-body types of vehicles the explosives shall be covered with a fire-resistant tarpaulin.

(2) All vehicles transporting explosives shall be marked with reflectorized placards on both sides and ends with the word EXPLOSIVES printed on.

(3) Blasting caps or other initiators shall not be transported in the same vehicle with other explosives, if possible; otherwise the caps should be carried in the front and the explosives in the rear of the truck.

(4) All vehicles used for transportation of explosives shall be in charge of, and operated by, a person who is mature-minded, physically fit, careful, reliable, able to read and write the English language, and not addicted to the use of intoxicants or narcotics. He should be aware of the destructive effects of explosives.

(5) No metal tools, carbides, oils, matches, firearms, electric storage batteries, flammable substances, acids, or oxidizing or corrosive compounds shall be carried in the bed or body of any vehicle transporting explosives.

(6) Vehicles to be used in the transportation of explosives shall be in good repair. When steel or part steel bodies are used, fire-resistant and nonsparking cushioning materials shall be employed to separate the containers of explosives from the metal.

(7) Vehicles transporting explosives shall be equipped with not less than two fire extinguishers placed at strategic points, filled and ready for immediate use, and of a make approved by the National Board of Fire Underwriters for class B and C fires.

(8) A vehicle containing explosives shall not be taken into a public building or repair shop or parked in congested areas for any period of time.

(9) All vehicles shall be checked before

transporting explosives and all electric wiring completely protected and securely fastened to prevent short circuiting.

(10) Vehicles transporting explosives shall be operated with extreme care and shall not be driven at a speed greater than 35 miles per hour. Full stops shall be made at approaches to all railroad crossings and main highways, and the vehicle shall not proceed until it is known that the way is clear. This, however, does not apply to convoys or protected crossings manned by highway flagmen or guards.

(11) All vehicles transporting explosives on public highways, roads, or streets shall have an authorized driver and helper. No person other than the authorized driver and helper shall be permitted to ride on trucks transporting explosives or detonators.

142. Magazines

a. Types. Explosives are stored in magazines according to the safety regulations prescribed in TM 9-1903. Table XV indicates the minimum distance for the location of magazines from other magazines, buildings, and routes of communication based on the quantity of explosives stored. There are two types of magazines—permanent and temporary. Although the permanent type is preferred, temporary or emergency types are frequently required when permanent construction is not possible.

b. Barricades. Explosives storage magazines must be barricaded, that is, have a substantial obstacle between them and inhabited buildings. For certain explosives, effective natural or artificial barricades reduce by one-half the distance necessary between magazines, railways, and highways. The use of barricades thus permits the storage of larger quantities of explosives in any given area. Although barricades help protect magazines against explosives and bomb or shell fragments, they are no safeguard against pressure damage.

c. Other Considerations. Magazines are usually placed at locations determined according to safety, accessibility, dryness, and drainage. Safety and accessibility, however, are the most important. An ideal location is a hilly area where the height of the ground above the magazine provides a natural wall or barrier to buildings, centers of communication, and other magazines in the area. Sidehill dugouts are not desirable, as adequate ventilation and drainage are often hard to provide. Brush and tall grass should be cleared from the site to minimize the danger of fire.

Table XV. Magazine Locations (Unbarricaded)

Qantity, pounds of explosives (not over)	Minimum distance in feet from nearest—		
	Inhabited building	Magazine	Public highway, railway, and/or electric lines
50	300	50	180
100	380	50	230
2,000	1010	140	610
20,000	1950	300	1170
100,000	3630	510	2180
225,000	4190	670	2515

Note. For more detailed information see TM 9-1903, C 1, 26 February 1957.

d. Lightning Protection. All magazines must have a grounded overhead lightning rod system. Also, all metal parts—doors, ventilators, window sashes, and reinforcing steel—must be connected in several places to buried conduits of copper-plate or graphite rods.

143. Field Expedient Structures

a. Field expedients for the storage of explosives when magazine construction is not possible are—

(1) A dugout excavated in a dry area and revetted with timber to prevent caving.

(2) An isolated building.

(3) A light wooden frame box house, with a wedge type roof covered by corrugated iron, or merely covered with a tent or canvas tarpaulin. The explosives should be placed on pallets for all-around ventilation.

b. Field expedient storage facilities should be appropriately marked by signs on all four sides, and guarded.

144. Temporary Magazines and Storage

Limited supplies of explosives can be stored for several days when necessary in covered ammunition shelters and should be so separated that fire or explosion cannot be transmitted from one shelter to another. Piles of explosives temporarily stored in the open should contain no more than 500 pounds, and be placed no less than 140 feet apart. Explosive components should be piled separately. Explosives, caps, and other demolition material stored temporarily in training areas should be kept separate in covered ammunition shelters, and under guard at all times. Temporary storage operations should be guided by local safety SOP's and other regulations (AR 385-63).

145. Destruction and Disposal of Explosives

a. Methods. Explosives, being insoluble in water, generally cannot be disposed of as sewage. Submergence, burning, or decomposition by chemical agents is necessary. Explosive material may be disposed of without alteration in form by dumping at sea. The best method of destroying explosives, however, is by burning.

b. Ordnance Units. Explosives are destroyed by explosive ordnance disposal units as directed in AR 75-15, TM 9-1385-9, TM 9-1900, TM 9-1903, and TM 9-1375-200.

APPENDIX A

REFERENCES

DA Pam 27–1	Treaties Covering Land Warfare.
AR 75–15	Responsibilities and Procedures for Explosive Ordnance Disposal.
AR 385–63	Regulations for Firing Ammunition for Training, Target Practice, and Combat.
SM 9–5–1375	Stock List of Current Issue Items–FSC Group 13 Ammunition and Explosives, Class 1375 Solid Propellants, and Explosive Devices.
FM 5–15	Field Fortifications.
FM 5–26	Employment of Atomic Demolition Munitions (ADM).
FM 5–29	Passage of Mass Obstacles.
FM 5–31	Boobytraps.
FM 5–34	Engineer Field Data.
FM 5–35	Engineer's Reference and Logistical Data.
FM 8–10	Medical Service, Theater of Operations.
FM 19–30	Physical Security.
FM 20–32	Land Mine Warfare.
FM 20–33	Combat Flame Operations.
FM 23-30	Grenades and Pyrotechnics.
FM 31–10	Barriers and Denial Operations.
FM 101–31–1	Staff Officer's Field Manual; Nuclear Weapons Employment.
TM 5-220	Passage of Obstacles Other than Minefields.
TM 5–280	Foreign Mine Warfare Equipment.
(C) TM 5-280A	Foreign Mine Warefare Equipment (U).
TM 5–332	Pits and Quarries.
TM 9–1300–206	Care, Handling, Preservation, and Destruction of Ammunition.
TM 9–1345–200	Land Mines.
TM 9–1375–200	Demolition Materials.
TM 9–1375–203–12	Operator and Organizational Maintenance · Manual: 10-Cap Capacity Handle Operated Blasting Machine.
TM 9–1385–9	Explosive Ordnance Reconnaissance.
TM 9–1900	Ammunition, General.
TM 9–1903	Care, Handling, Preservation, and Destruction of Ammunition.
TM 9–1910	Military Explosives.
GTA 5–10–9	Demolition Card.

APPENDIX B

METRIC CHARGE CALCULATIONS

1. Introduction

The following metric equivalent charge calculation formulas are included because of NATO requirements, wherein the United States and British Armies are gradually changing over from their tables of measurement to the metric system. Problems, solutions, and tables with measurements converted to the metric system are given below.

2. Structural Steel Cutting Formula

Formula: $K = \frac{A}{38}$

K = kilograms of TNT required

A = gross sectional area in square centimeters

Example: (fig. 97)

Flange Area = 2 × 1.2 × 12.7 = 30.48 or 30.5 sq cm

Web Area = 28 × 1 = 28 sq cm

A (total) = 30.5 sq cm + 28 sq cm = 58.5 sq cm

$K = \frac{A}{38} - \frac{58}{38}$

K = 1.5 (use 1.5 kilograms of TNT)

3. Timber Cutting Formula

a. External Charge.

Formula: $K = \frac{D^2}{550}$

K = kilograms of TNT required

D = diameter of target in centimeters

Example: the diameter of a tree is 30 centimeters.

$K = \frac{30^2}{550} = \frac{900}{550}$

K = 1.64 kg

Use of 1.6 kilograms of TNT

b. Internal Charge.

Formula: $K = \frac{D^2}{3500}$

K = kilograms of TNT required

D = diameter of target in centimeters

Example: (fig. 96)

$K = \frac{D^2}{3500}$

$K = \frac{30^2}{3500} = \frac{900}{3500}$

K = .257 kg

Use 260 grams of TNT or any other explosive

4. Breaching Formula

Formula: $K = 16R^3KC$

K = kilograms of TNT required.

R = breaching radius in meters.

K = the material factor based on strength and hardness of material to be demolished.

C = the amping factor based on type and extent of tamping to be used.

Add 10 percent to calculated charge less than 22.5 kilograms.

For walls 30 centimeters (approx 1 ft) thick or less, increase the charge by 50 percent.

Table XVI. Value of Material Factor K for Calculation of Breaching Charges (Metric)

Material	Breaching radius	K
Ordinary earth	All values	0.06
Poor masonry, shale and hardpan, good timber, and earth construction	All values	0.28

Material	Breaching radius	K
Good masonry, ordinary concrete, rock	Less than 1 meter	0.35
	1 to less than 1.5 meters	.28
	1.5 to less than 2 meters	.25
	More than 2 meters	.23
Dense concrete, first-class masonry	Less than 1 meter	0.45
	1 to less than 1.5 meters	.38
	1.5 to less than 2 meters	.33
	2 or more meters	.28
Reinforced concrete (will not cut steel reinforcing)	Less than 1 meter	0.70
	1 to less than 1.5 meters	.55
	1.5 to less than 2 meters	.50
	2 or more meters	.43

a. Breaching Radius. The breaching radius (R) is the distance in meters which an explosive charge must penetrate and within which all material is displaced or destroyed. For example, to breach a 2-meter concrete wall by placing a charge on one side, the value of R in the formula $K = 16 R^3KC$ is 2 meters.

b. Material Factor. The values of the material factor for various types of construction are given in table XVI.

c. Tamping Factor. The value of the tamping factor depends on the location and the tamping of the charge. No charge is considered fully tamped unless it is covered to a depth equal to the breaching radius. If underwater demolition is necessary, the tamping factor for placement of charges tamped with earth is used (fig. 105).

Example: Determine the amount of TNT required to breach a dense concrete pier 1.5 meters thick with untamped charges placed on the ground.

$R = 1.5$ meters

$K = .33$ (dense concrete, table XVI)

$C = 4.5$ (untamped, on the ground, fig. 105)

$K = 16R^3KC$

$K = 16 \times 3.3 \times .33 \times 4.5$

$K = 78.4$ or 78.5 kilograms of TNT per charge

5. Additional Data

Characteristics of U.S. explosives, steel cutting charges, and minimum safe distances, expressed in the metric system, are given in tables XVII, XVIII, and XIX respectively.

Table XVII. Characteristics of Principal U. S. Explosives (Metric)

Name	Velocity of detonation (meters per sec)	Relative effectiveness	Weight per block
TNT	6,900	1.00	.454 and .227 kg
Tetrytol	7,000	1.20	1.1 kg
M118 (sheet explosive).	7,190		.90 kg—block
Composition C-3.	7,625	1.34	0.22 kg—sheet 1.02 kg—M3
Composition C-4.	8,040	1.34	1.1 kg—M5 1.1 kg—M5A1
Ammonium nitrate.	3,400	0.42	0.56 kg—M112 18.17 kg
Military dynamite.	6,100	0.92	.227 kg

Table XVIII. Steel Cutting Charges (Metric)

Average thickness of section in centimeters	Kilograms of TNT for rectangular steel section of given dimensions												
	Width of Sections in Centimeters												
	5.0	7.6	10.2	12.7	15.2	20.3	25.4	30.5	35.6	40.6	45.7	50.8	61.0
.64	.09	.14	.18	.23	.27	.36	.45	.54	.64	.73	.82	.91	1.1
.95	.14	.23	.27	.32	.41	.54	.64	.77	.91	1.04	1.18	1.27	1.54
1.27	.18	.27	.36	.45	.54	.68	.86	1.04	1.22	1.36	1.5	1.7	2.0
1.59	.23	.32	.45	.54	.64	.86	1.1	1.32	1.5	1.7	1.95	2.13	2.59
1.91	.27	.41	.54	.64	.77	1.00	1.27	1.54	1.81	2.04	2.31	2.59	3.08
2.22	.32	.45	.64	.77	.90	1.22	1.5	1.81	2.08	2.4	2.7	3.0	3.6
2.54	.36	.54	.68	.86	1.04	1.36	1.7	2.0	2.4	2.7	3.1	3.4	4.1

Table XIX. Minimum Safe Distances (in the Open) (Metric)

Kilos of explosives	Safe distance in meters	Kilos of explosives	Safe distance in meters
.45 to 12 kilos	274	84	384
13	281	86	393
14	290	40	410
16	302	45	424
18	311	56	457
20	320	62	486
22	329	90	534
23	337	136	612
25	348	181	673
27	357	226	726
29	366	Over 226 (compute by formula)	

APPENDIX C

USE OF LAND MINES, AERIAL BOMBS, AND SHELLS AS DEMOLITION CHARGES

1. Introduction

When land mines, aerial bombs, and shells are used as demolition charges, special precautions must be taken because of flying steel fragments. The use of such mines, bombs, and shells is generally uneconomical but may at times become necessary or desirable. Such material may be issued from captured or friendly supply stocks or, in the case of land mines, may be those recovered from enemy or friendly minefields. *In no case should unexploded dud shells or bombs be used for demolition purposes.*

2. Land Mines

a. Safety Precautions. Only defuzed mines should be used in demolition charges, as fused mines recovered from minefields may be sensitive because of near misses and may be detonated by even normal handling. The use of enemy mines salvaged from minefields or dumps is regulated by directives issued from headquarters of the theater concerned. United States and foreign land mines are described in detail in TM 9-1345-200, TM 5-280, and TM 5-280A.

b. Charges. In calculating demolition charges when using mines, only the explosive weight is considered. Normal explosive quantities may be used for cratering or pressure charges with mines; but, because of poor contact of the mine case against irregularly shaped objects, it may be necessary to increase cutting charges considerably. Test shots will determine the results to be obtained under given conditions. A list of antitank mines in current use by the United States and (in current use or obsolete in foreign armies) with their explosive weights is given below. Information, however, on the type of explosive used is not always available—

(1) *United States*

Type Mine	*Explosive*
M7A2 A/T (metallic)	3⅝ lb TNT
M6A2 A/T (metallic)	12 lb TNT
M15 A/T (metallic)	22 lb TNT
M19 A/T (nonmetallic)	21 lb TNT
M21 A/T (metallic)	10½ lb composition H6

(2) *Foreign Mines*

(*a*) *Austria*

Barrier A/T (metallic)	10 lb

(*b*) *Belgium*

Model VI A/T (metallic)	6 lb
BSB A/T (metallic)	7.75 lb TNT
Type H A/T (metallic)	12.75 lb TNT
Type HA A/T (metallic)	12.75 lb TNT

(*c*) *Communist China*

Dual purpose No 8 (metallic).	5 lb
Dual purpose No 4 (metallic).	12 lb
Model 1951 A/T (wooden)	13.8 lb TNT

(*d*) *Czechoslovakia*

PT-Mi-K A/T (metallic)	11 lb TNT
PT-Mi-D A/T (wooden)	6 lb TNT (approx)
PT-Mi-Ba A/T (plastic or bakelite).	15 lb cast TNT

(*e*) *Finland*

M 36 A/T (metallic)	8 lb TNT
M 39 A/T (metallic)	8.8 lb TNT

(*f*) *France*

M1935 heavy A/T (metallic).	3.25 lb
M1936 light A/T (metallic).	5.75 lb
M1948 A/T (metallic)	11.5 TNT or MD (20% dinitronapthalene and 80% picric acid*)
M1948 plate charge A/T (metallic).	16 lb TNT or picric acid*
M1951 shaped charge A/T (metallic).	4 to 5 lb hexolite
Model 1947 A/T (plastic)	12.1 lb TNT
Model 1951 A/T (Caseless)	16–20 lb cast TNT

(*g*) *Hungary*

CVP-1 variable pressure, general purpose (metallic).	3.5 lb TNT

See footnote on page 158.

Type Mine	*Explosive*
(h) Japan (WW II)	
Type 93 antivehicular (metallic).	2 lb picric acid*
Yardstick antivehicular (metallic).	6 lb picric acid*
Model 1 beach mine (double horn) (metallic).	40.6 lb trinitroanisol
Model 2 beach mine (single horn) (metallic).	22 lb
(i) Netherlands	
Type II A/T (metallic)	9 lb
Mushroom-topped dual purpose (metallic).	5.25 lb TNT
(j) South Korea	
Heavy A/T mine (metallic)	22 lb TNT
Type I dual purpose (metallic).	5.7 lb flaked TNT
Type II dual purpose (metallic).	4.5 lb TNT
(k) USSR	
PMZ-40 A/T (metallic)	8 lb
TM-35 A/T (metallic)	8.8 lb
TM-38 A/T (metallic)	6.5 lb
TM-41 A/T (metallic)	8 lb amatol 80/20 or flaked TNT (picric acid* booster)
T-IV A/T (metallic)	6.2 lb
AKS general purpose (metallic).	13.2 lb
TMD-B A/T (wooden)	11-15 lb pressed amatol, dynammonite, cast TNT, or powdered picric acid*
TMB-44 A/T (wooden)	11-15.4 lb amatol, dynammonite, or TNT
TMB-2A/T (tar-impregnated cardboard).	11 lb powdered amatol 80/20
TMS-B A/T (tar-impregnated cardboard).	13 lb powdered amatol 80/20
(l) United Kingdom	
Mark 2 EP A/T (metallic)	4.5 lb TNT
Mark 2 GS A/T (metallic)	4 lb TNT or baratol (barium nitrate and TNT-20/80 or 10/90)
Mark 3 GS A/T (metallic)	4.5 lb TNT
Mark 4 GS A/T (metallic)	8.25 lb TNT
Mark 5 GS A/T (metallic)	4.5 lb TNT
Mark 5 HC A/T (metallic)	8.3 lb TNT
Mark 6 EP and Mark 5 C A/T (metallic)	4.5 TNT

* *Picric acid corrodes metals, forming extremely sensitive compounds, easily detonated. Mines loaded with this explosive should not be handled except to move them to a safe disposal area for destruction.*

c. Priming. Land mines are detonated by means of a pound of explosive placed on the pressure plate. If large quantities of mines are to be fired simultaneously, several mines are primed to insure complete detonation. Detonation of a single mine normally detonates other mines in contact with it.

3. Aerial Bombs

a. Use. General-purpose aerial bombs may be used satisfactorily as demolition charges but are more effective as cratering charges. Their shape makes them inefficient for demolitions requiring close contact between the explosive and the target. Precautions must be taken to avoid damage to installations and injury to personnel because steel fragments of the bomb case are thrown great distances. Before using a bomb, it must be positively identified as a general-purpose bomb.

b. Charges. The explosive content of bombs is approximately half their total weight. Table XX gives the weight of high explosive in various types of general-purpose bombs. Approximately 20 percent of the explosive power is expended in shattering the case.

Table XX. Explosive Content of General-Purpose Bombs

Bomb	Total weight (lb)	Explosive weight (lb)
100-lb GP, AN-M30A1	115	57
250-lb GP, AN-M57A1	260	125
500-lb GP, AN-M64A1	525	266
1,000-lb GP, AN-M65A1	990	555
2,000-lb GP, AN-M66A2	2,100	1,098
3,000-lb GP, T-55	2,605	1,710

c. Priming. Bombs under 500 pounds weight are detonated by firing a 5-pound explosive charge in good contact in the middle of the case. Bombs of 500 pounds or more are detonated by a 10-pound charge similarly placed. Fuses should not be positioned on the nose or tail. To insure detonation, large bombs should be primed separately.

4. Artillery Shells (Nonnuclear)

Artillery shells are used for demolition only where a fragmentation effect is desired. Because of their low explosive content they are seldom used for other demolition purposes. The

105-mm howitzer HE shell, which weighs 33 pounds, contains only 5 pounds of explosive; while the 155-mm howitzer shell contains only 15 pounds. Shells up to 240-mm are detonated by 2 pounds of explosive placed in good contact with the case, just forward of the rotative band. To insure complete detonation, a charge should be placed on each shell. The universal destructor M10 (para 40*a*) may be used to detonate projectiles or bombs that have 1.7- or 2-inch diameter threaded fuse wells. The booster cavities of bombs and large projectiles should be filled to the full depth by adding booster cups to the destructor M10 as required.

APPENDIX D

SUMMARY OF EXPLOSIVE CALCULATION FORMULAS

1. Timber-Cutting Charges

a. External Charges, Untamped (para 78*a*).

$P = \frac{D^2}{40}$

P = pounds of TNT required

D = diameter of the timber in inches or the least dimension of dressed timber.

b. Cutting Trees to Create an Obstacle (para 78*b*).

c. Internal Charges, Tamped (para 78*c*).

$P = \frac{D^2}{50}$

P = pounds of any explosive

D = diameter, or the least cross-sectional dimensional in inches.

2. Steel-Cutting Charges

a. Structural Members (para 81*b*(1)).

$P = \frac{3}{8} A$

P = pounds of TNT required

D = cross-sectional area in square inches of the steel member to be cut.

b. Other Steel Members (para 81*b*(2)(*a*)).

$P = D^2$

P = pounds of TNT required

D = diameter, in inches, of section to be cut.

c. Steel Bars 2 Inches in Diameter or Less (para 81*b*(2)(*b*)).

$P = D$

P = pounds of TNT required

D = diameter of bar in inches or largest dimension of section to be cut.

Rule of thumb.

Bars up to 1 inch in diameter, use 1 pound TNT.

Bars over 1 inch in diameter and up to 2 inches, use 2 pounds TNT.

d. Railroad Rails.

To cut 80-pound or lighter rail (5 inches or less in height), use ½ pound of explosive

To cut rails over 80 pounds (over 5 inches high), use 1 pound of explosive.

e. Saddle Charge (para 83*b*).

Base of charge = ½ circumference of target (fig. 100).

Long axis of charge = circumference of target

Thickness of charge = 1/3 thickness of M5A1 block (2/3 inch) for targets up to 19 inches in circumference (6 inches in diameter); ½ the thickness of M5A1 block (1 inch) for targets from 19 to 25 inches in circumference (over 6 to 8 inches in diameter).

Note. Steel alloy targets over 25 inches in circumference (over 8 inches in diameter) require the diamond charge.

f. Diamond Charge (para 83*c*).

Long axis of charge = circumference of target (fig. 101).

Short axis of charge = ½ circumference of target

Thickness of charge = 1/3 thickness M5A1 block (2/3 inch)

g. Ribbon Charge (para 83*d*).

Thickness of charge = ¾ thickness of target (fig. 102)

Width of charge = 3 × thickness of charge

Length of charge = length of cut.

3. Pressure Charges

(para 84)

$P = 3H^2T$

P = pounds of TNT required for each stringer

H = height of stringer, including thickness of roadway

T = thickness in feet of stringer in feet

The values of H and T, if not whole numbers, are rounded off to the next higher quarter-foot dimension. Neither H nor T is ever considered to be less than 1 in the formula.

Note. Increase the calculated charge P by one-third if it is not tamped.

4. Breaching Charges

a. Size of Each Charge (para 86*a*).

$P = R^3KC$

P = pounds of TNT required.

R = breaching radius in feet (rounded off to the next higher ½-foot).

K = material factor (table XI).

C = tamping factor (fig. 105).

Note. Add 10 percent to the calculated charge whenever P is less than 50 pounds and increase the charge by 50 percent for walls 1 foot thick or less.

b. Number of Charges (para 87*b*).

$$N = \frac{W}{2R}$$

N = number of charges

W = width of pier, slab, or wall in feet

R = breaching radius in feet

When the value of N has a fraction less than ½, the fraction is disregarded, but when the fraction is ½ or more, the value is rounded off to the next higher whole number. An exception to the general rule is the N-value between 1 and 2, wherein a fraction less than ¼ is disregarded, but a fraction of ¼ or more is rounded off to the next higher whole number, 2.

5. Cratering Charges

a. Deliberate Method (para 89). Forty-pound charges in 6-foot boreholes are alternated with 80-pound charges in 7-foot boreholes. All boreholes are placed on 5-foot centers. The end holes in all cases are 7 feet deep. No two 5-foot holes should be adjacent to each other (fig. 107).

b. Hasty Method (para 90). Ten pounds of explosive per foot of borehole is placed in holes of equal depth. Boreholes are positioned on 5-foot centers at depths varying from 2½ to 5 feet (fig. 108).

c. Relieved Face Crater (para 91*a*). Two rows of boreholes are drilled 8 feet apart (fig. 109); boreholes are spaced on 7-foot centers—four on the friendly side and three staggered between them on the enemy side. Boreholes on friendly side are 5 feet deep and loaded with 40 pounds of explosive, and on the enemy side, 4 feet deep and loaded with 30 pounds of explosive. Row on enemy side is detonated first and on the friendly side, ½ to 1½ seconds later.

d. Angled Crater (para 91*b*). A line of boreholes is blasted or drilled across a roadway at a 45° angle (fig. 110). Standoff distance for M2A3 shaped charge for boring holes on unpaved roads is from 20 to 30 inches; and on paved roads, about 36 inches. Increase in standoff distance increases depth of borehole but decreases its diameter.

6. Breaching Hard-Surfaces Pavements

Charges are computed on the basis of 1 pound of explosive per 2 inches of pavement thickness. Tamping should be twice the thickness of the pavement (para 88*b*(2)).

7. Computation of Minimum Safe Distances

a. For charges less than 28 pounds, the minimum safe distance is 900 feet. This, however, gives no insurance against missile hazards, which require a defilade.

b. For charges from 28 to 500 pounds, the safe distance is computed by means of this formula:

Safe distance in feet =

$$300 \sqrt[3]{\text{pounds of explosive}}$$

c. For quarrying operations the formula is:

Safe distance in feet =

$$350 \sqrt[3]{\text{pounds of explosive}}$$

8. Notes

a. The charges calculated by the above formulas should be rounded off to the next higher unit package of explosive being used or cut, when applicable.

b. When an explosive other than TNT is used in external charges computed from the steel, timber, breaching, or pressure formula, the value of P should be adjusted by use of the relative effectiveness factor as indicated in table VIII.

APPENDIX E

POWER REQUIREMENTS FOR SERIES FIRING CIRCUIT

1. Series Circuit

In demolition projects, electric blasting caps are connected in series and fired by an electric power source (blasting machine). A series circuit provides a single path for the electrical current which flows from one firing wire through each blasting cap to the next blasting cap and back to the other firing wire. A series circuit should not contain more than 50 blasting caps. The connection of more than 50 caps in a series circuit increases the hazard of breaks in the firing line or cap leads prior to the initiation of some caps.

2. Ohm's Law

The amount of voltage necessary to detonate the blasting caps in these circuits is calculated by the use of the basic law of electricity, Ohm's Law—

$E = IR$

E = electrical potential, or voltage, expressed in volts.

I = current, expressed in amperes.

R = resistance, expressed in ohms.

3. Electric Power Formula

Electrical power is computed by means of the following formula:

$W = I^2R$

W = electrical power, expressed in watts.

I = current, expressed in amperes.

R = resistance, expressed in ohms.

4. Electrical Characteristics of Electric Blasting Caps

The current needed to fire military electric blasting caps connected in series should be at least 1.5 amperes regardless of the number of caps. The resistance of a military electric blasting cap is 2 ohms.

5. Resistance of a Circuit

Resistance is computed to insure that the power source is adequate to fire all charges connected to the circuit. Both the blasting caps and the wire contained in a circuit contribute to the total resistance of that circuit. This resistance is computed from the individual resistances of the blasting caps and the wire.

Table XXI. Resistance of Various Sizes of Copper Wire

1	2	3	4
Size o copper wire			
AWG (B&S) gage No.	Diameter (in.)	Length of wire to weigh 1 pound (ft. per lb.)	Resistance of 1,000 feet of wire (ohms per 1,000 ft.)
2	3/10	5.0	0.2
4	1/4	7.9	.3
6	1/6	12.6	.4
8	1/8	20.0	.6
10	1/10	31.8	1.0
12	1/12	50	1.6
14	1/16	80	2.5
16	1/20	128	4.0
18	1/25	203	6.4
20	1/30	323	10.2

The resistance of the wire used is a circuit depends upon its size and the length. Table XXI gives the resistance per 1,000 feet of various sizes of copper wire. The total resistance in a series circuit is the sum of the resistance of the various components of that circuit. (For simplicity of calculation in the field, only the resistance of the blasting caps is used to determine the resistance of a circuit.)

6. Calculations for a Series Circuit

Complete calculations for any circuit involve the determination of the current (amperes), the voltage (volts), and the power (watts) needed to fire the circuit. Computation of the voltage and of the power requires the determination of the resistance (ohms) in the system.

a. Current Requirements. The current required for a series-connected system of special electric blasting caps is 1.5 amperes, regardless of the number of blasting caps in the circuit.

b. Resistance. The resistance of the system is computed as described in paragraph 5 of this appendix.

c. Voltage Requirements. Using Ohm's Law, $E = IR$ (para 2 this app), the voltage needed is computed by multiplying the required current (1.5 amperes) by the resistance of the system.

d. Power Requirements. By means of the electrical power formula, $W = I^2R$ (para 3 this app), the number of watts of power needed may be found by multiplying the square of the current required ($1.5^2 = 2.25$) by the resistance of the system.

e. Illustrative Problem. Determine the current, voltage, and power required to detonate the blasting caps of a circuit consisting of 20 special electric blasting caps connected in series, and 500 feet of the standard 2-conductor, 18-gage firing wire.

(1) Current required = 1.5 amperes (*a* above)

(2) Resistance:

20 blasting caps = $2.0 \times 20 = 40$
1,000 feet No. 18 wire (table XXI)
6.4
Total resistance = 46.4 ohms

Note. As 500-foot firing wire consists of 2 strands of No. 18 wire each 500 feet long, 1,000 feet of wire is used in the above computation.

(3) Voltage:

$E = IR$ (para 2 this app)
$E = 1.5 \times 46.4 = 69.6$ volts

(4) Power:

$W = I^2R$ (para 3 this app)
$E = 1.5^2 \times 46.4 = 2.25 \times 46.4 =$
$E = 1.5^2 \times 46.4 = 2.25 \times 46.4$
104.4 watts

7. Calculated Voltage Drop

In each of the examples given above the voltage drop (IR) in the blasting circuit was calculated by the use of Ohm's Law. In practice, if the calculated voltage drop exceeds 90 percent of the available voltage, it is recommended that the resistance of the circuit be decreased or the voltage be increased.

8. Capacity of Power Sources

a. Determining Capacity of Power Sources. It is possible to determine from the nameplate amperage and voltage rating whether the power source is suitable for firing an electric circuit computed by the above methods. Frequently, however, the size of a circuit that may be fired with current from a given power source may be determined by consulting table XXII which gives the maximum capacities of some power sources. If it is necessary to calculate the capacity of a given generator from the nameplate data, proceed as follows:

(1) Divide 90 percent of the voltage of the generator (para 6 this app) by the total amperage of the circuit, 1.5 amps, to determine the maximum resistance in ohms that may be in the circuit.

(2) Subtract the total wire resistance from the maximum allowable circuit resistance of caps to determine the maximum allowable resistance of the caps in the circuit.

(3) To calculate the maximum number of caps, divide the allowable resistance of the caps in the circuit by the resistance of one cap (2.0 ohms).

b. Illustrative Problem. Determine the number of military electric blasting caps in series that may be fired by a 220-volt, 13½

Table XXII. Maximum Circuit Capacities of Various Power Sources

1	2	3	4	5	6	7	8	9	10
Circuit design	Total number of caps in circuit	Power source							
		10-cap blasting machine	30-cap blasting machine	50-cap blasting machine	1½-kw portable generator, 115-volt, 13½-amp	3-kw portable generator, 115-volt, 26-amp	5-kw portable generator, 115-volt, 43½-amp	3-kw portable generator, 220-volt, 13½-amp	5-kw portable generator, 220-volt, 22½-amp
		The circuits below are connected by one 500-foot standard two-conductor firing reel							
1 10 caps in continuous series.	10	x	x	x	x	x	x	x	x
2 30 caps in continuous series.	30	-----	x	x	x	x	x	x	x
3 50 caps in continuous series.	50	-----	-----	x	-----	-----	-----	x	x

ampere generator using 500 feet of 20-gage connecting wire.

(1) Allowable resistance of circuit = $\frac{(0.90)(220)}{(1.5)}$ = 132 ohms (para 6 and 7).

(2) Resistance of firing wire = $\frac{(10.2)(500)}{1{,}000}$ = 5.1 ohms (table XXI)

(3) Allowable resistance of caps for a series circuit = 132 − 5.1 = 126.9 ohms

(4) Number of blasting caps allowed in the series circuit = $\frac{126.9}{2.0}$ = 63.4 or 63 caps (*a*(3) above).

c. Use of Storage Batteries and Dry Cells. The size of a circuit that may be fired by a battery or dry cell may be determined by following the same procedure as that outlined in *a*(1) through (3) above.

Caution: **For safety, disconnect the battery terminal prior to disassembly of the equipment where there is danger from shorting across the battery circuit. In reassembly, make the battery terminal connection last.**

APPENDIX F

SPECIAL DEMOLITION MATERIALS AND TECHNIQUES

Section I. SPECIAL CHARGES

1. Square Charge

a. Description. This technique is applicable to the demolition of concrete and masonry bridge piers and other types of construction, but not steel. The charge for use on reinforced concrete walls up to 4 feet thick is composed of composition C 4 blocks, 2 x 2 x 11 inches. They are placed as removed from the packing case. For walls from 5 up to 7 feet thick, haversacks of 8 blocks of C 4 (M5A1) explosive, measuring 4 x 8 x 11 inches each may be used. The blocks are not removed from the haversacks, as they are easily fastened against the target (fig. 121). The size of the charge depends on the thickness of the target and the ratio of the thickness of the charge and the contact area. Although these charges, if square, are more effective than if rectangular, it is not always feasible to cut them to size. As most charges are rectangular, additional explosive is allowed for modification in technique. The charges tabulated in *c* below, have proved effective.

b. Placement on Piers.

(1) Place the charge at least the thickness of the target above the base to obtain the maximum results. A small charge may be taped to the target or supported by a platform. Larger charges may be supported by strips of material and wire attached to the pier by fasteners driven into the concrete by means of the powder-actuated driver.

(2) Initiate the charge from the center (fig. 121).

(3) Mud tamp the explosive on 1-foot thick targets, as this permits a 30-percent reduction in explosive weight.

c. Charge Size.

Concrete thickness	Charge size	Charge thickness
1 ft	2 C4 blocks	One block—2 in
2 ft	4 C4 blocks	One block—2 in
3 ft	7 C4 blocks	One block—2 in
4 ft	20 C4 blocks	One block—2 in
5 ft	6 M37 kits (20 lb packet)	One kit—4 in
6 ft	8 M37 kits (20 lb packet)	One kit—4 in
7 ft	12 M37 kits (20 lb packet)	One kit—4 in

2. Foxhole Digger Explosive Kit

a. Characteristics (fig. 141)

(1) Case

Material	Shape	Size	Wt
Plastic with screw cap.	Tubular with truncated top.	7.38 x 2.28 in	1.0 lb

(2) Shaped Charge

Material	Shape	Size	Explosive Charge			
			Type	Det Vel	Wt	Booster
Copper cone with 60° angle; and plastic.	Tubular with truncated top.	7.37 x 2.0 in	Octol	27,559 fps	118g (4.16 oz)	RDX

(3) Cratering Charge

Material	Shape	Size	Explosive Charge			
			Type	Det Vel	Wt.	Booster
2 segments of pressed explosive; connecting sleeve.	Tubular	8.21 x 1.0 in	PBXN-1	24,606 fps	162g (5.71 oz)	RDX

(4) Fuzes

Material	Shape	Size	Action	Initiation	Safety	Explosive
Stainless steel body; steel coupling.	Tubular	4.25 x 0.56 in	Mechanical with spring-driven striker	Push button	Cotter pin	RDX and primer.

(5) Auxiliary items

Piece of No. 9 nylon twine 36 in long; steel stability rod 4.25 x 0.1 in; two strips adhesive-coated foam tape; and lug on side of case with hole for stability rod and a ring for attaching kit to soldier's clothing or equipment

(6) Remarks

Outer case of kit serves also as standoff for shaped charge.

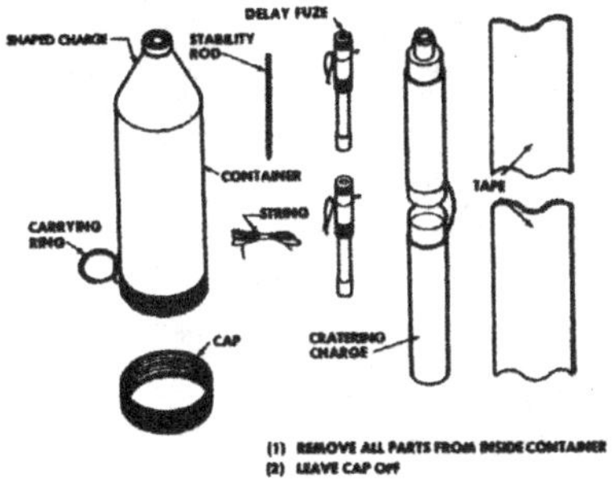

Figure 141. Foxhole digger explosive kit.

b. Effect.

(1) The shaped charge will penetrate soil, depending on the density, to depths varying from 20 to 33 inches, forming a tapered hole 2¼ inches in diameter at the top and 1 inch at the bottom. It will bore a hole through 5-inch mild steel plate, 1 inch in diameter at the entry and ¼ inch at the exit, and will penetrate concrete to a depth of 8

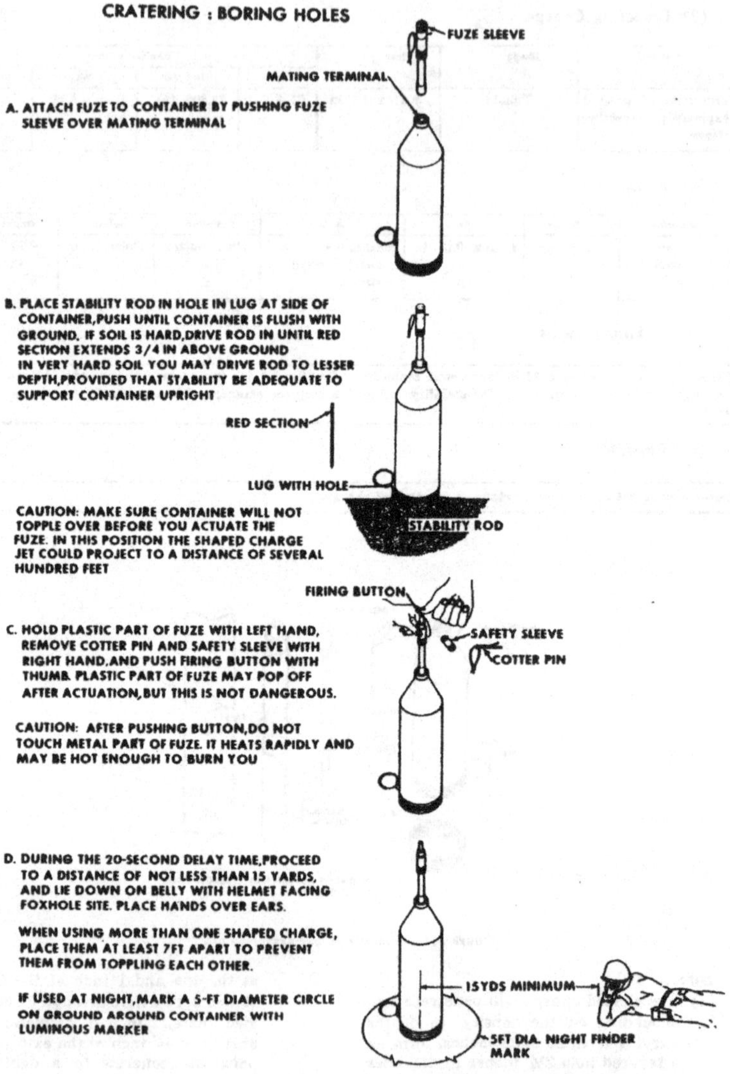

Figure 142. Arming and placement of foxhole digger explosive kit for cratering.

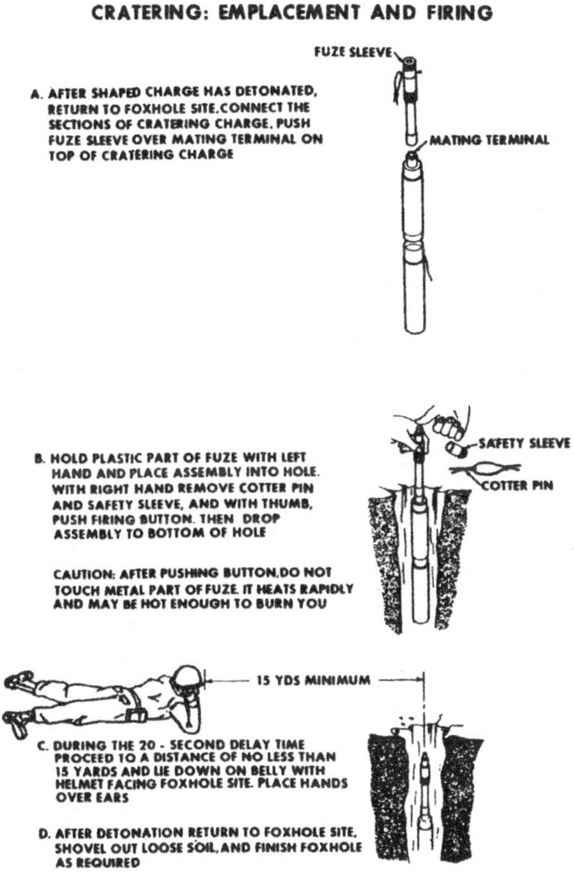

Figure 142—Continued.

inches with a hole 1¼ inches in diameter at the top and ½ inch at the bottom.

(2) The cratering charge will form a crater in soil about 42 inches in diameter and about 32 inches deep.

c. *Use*. For demolition purposes the shaped charge (5.71 ounces of explosive) may be useful in boring small holes in metal, concrete, wood, and soil, and in cutting small steel bars, rods, and cables. This, of course, depends on the ingenuity and initiative of the experienced demolitionist. It may be useful to damage metal working parts of vehicles and other

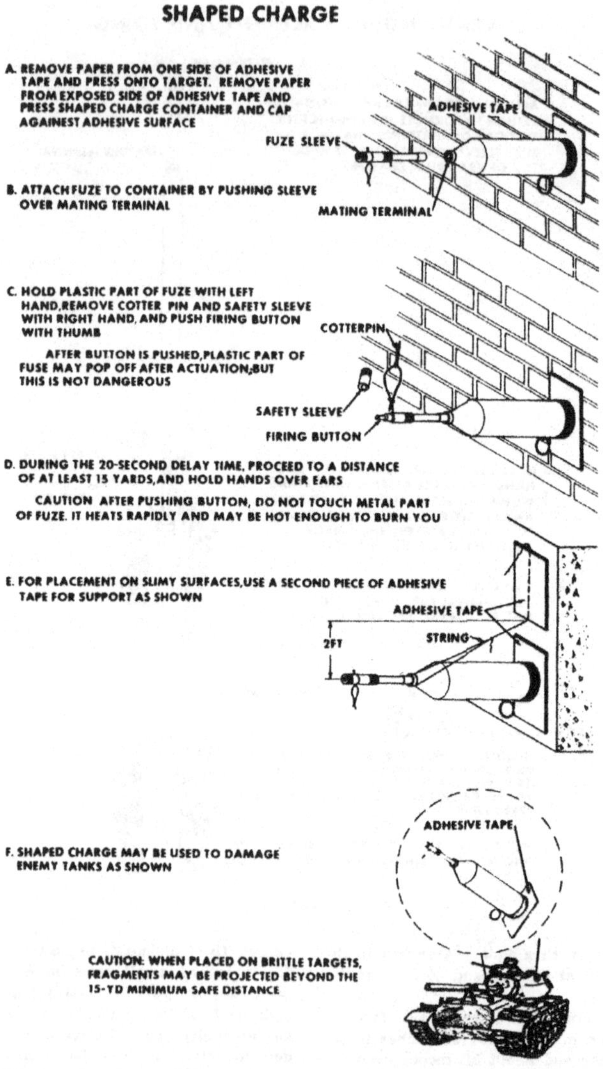

Figure 143. Arming and placement of foxhole digger explosive kit for damage.

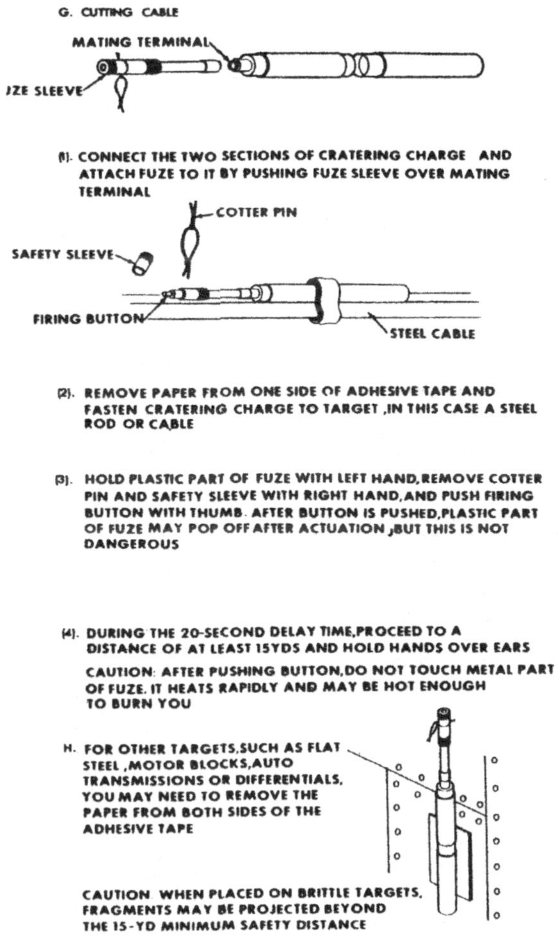

Figure 143—Continued.

equipment. The cratering charge, also being a high explosive (weighs 5.7 ounces) may be useful in cutting small metal bars, rods, and cables and in damaging equipment. Under critical conditions, however, test shots should be made to ascertain the effectiveness of the shaped charge and cratering charge.

d. *Arming and Placement.*

(1) For arming procedure and placement for cratering see figure 142.

(2) For arming procedure and placement for damage, see figure 143.

Note. As the delay period of the fuze may vary from 20 to 50 seconds between units, *users should consider the delay as 20 seconds for safety reasons.*

Section II. EXPEDIENT DEMOLITIONS

3. Use of Expedient Techniques

These techniques are not presented as a replacement for the standard demolition methods but for use by experienced blasters in special projects. Availability of trained men, time, and material will generally determine their use.

4. Shaped Charges

a. *Description.* Shaped charges concentrate the energy of the explosion released on a small area, making a tubular or linear fracture in the target. Their versatility and simplicity make them effective against many targets, especially those made of concrete or those with armor plating. Shaped charges may be improvised (fig. 144). Because of the many variables, such as explosive density, configuration, and density of the cavity liner, consistent results are impossible to obtain. Thus experiment, or trial and error, is necessary to determine the optimum standoff distances. Plastic explosive is best suited for this type of charge. Dynamite and molten TNT, however, may be used as an expedient.

b. *Preparation.* Almost any kind of container is usable (fig. 144). Bowls, funnels, cone-shaped glasses (champagne glasses with the stem removed), and copper, tin, or zinc may be used as cavity liners; or wine bottles with a cone in the bottom (champagne or cognac bottles) are excellent. If none of these is available, a reduced effect is obtained by cutting a cavity into a plastic explosive block. Optimum shaped charge characteristics are—

(1) Angle of cavity = between 30° and 60° (most HEAT ammunition has a 42° to 45° angle)

(2) Standoff distance = 1½ × diameter of cone.

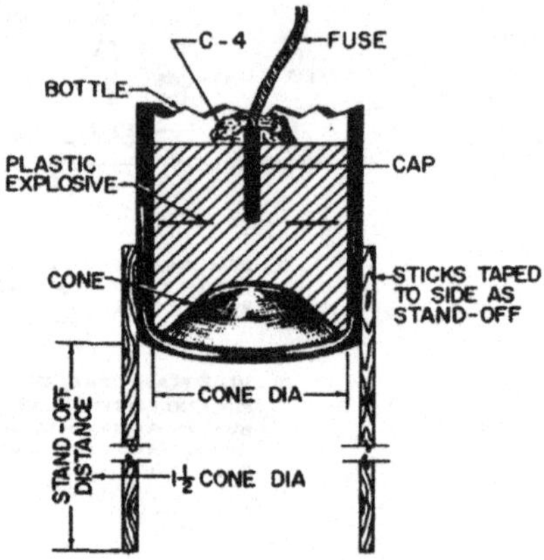

Figure 144. Improvised shaped charge.

(3) Height of explosive in container = 2 × height of cone measured from the base of the cone to the top of the explosive.

(4) Point of detonation = exact top center of charge. Cover cap, if any part of it is exposed or extends above the charge, with a small quantity of C4 explosive (fig. 144).

Note. The narrow necks of bottles or the stems of glasses may be cut by wrapping them with a piece of soft absorbent type twine or string soaked in gasoline and lighting it. Two bands of adhesive tape, one on each side of the twine or string, will hold it firmly in place. The bottle or stem must be turned continuously with the neck up, to heat the glass uniformly. Also, a narrow band of plastic explosive placed around the neck and burned gives the same result. After the twine or plastic has burned, submerge the neck of the bottle in water and tap it against some object to break it off. *Tape the sharp edge of the bottle to prevent cutting hands while tamping the explosive in place.*

5. Opposed (Counterforce) Charge

This technique is very effective against comparatively small cubical concrete and masonry objects 4 feet or less in thickness. If properly constructed of plastic explosive, placed, and detonated, counterforce charges produce excellent results with a relatively small amount of explosive. Their effectiveness results from the simultaneous detonation of two charges placed directly opposite each other and as near the center of the target as possible (fig. 145).

a. Charge Calculation. The size is computed from the diameter or thickness of the target in feet, as—

> The amount of explosive = 1½ × the thickness of the target in feet (1½ pounds per foot).

Fractional measurements are rounded off to the next higher foot prior to multiplication. For example, a concrete target measuring 3 feet 9 inches thick requires 1½ × 4 = 6 pounds of plastic explosive.

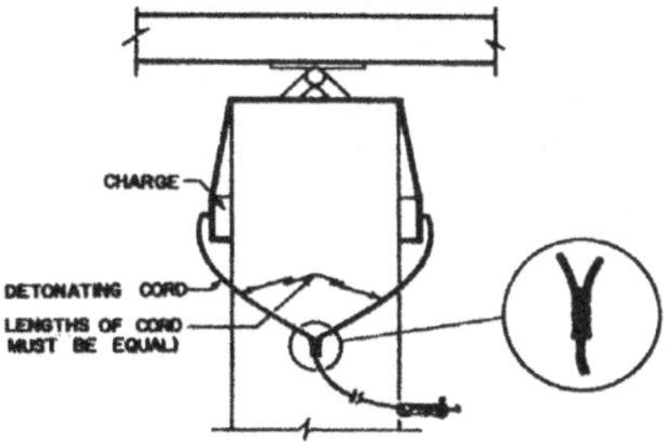

Figure 145. Opposed charge.

d. Preparation and Emplacement. Divide the calculated amount of explosive in half to make two identical charges. The two charges *must* be placed diametrically opposite each other. This requires accessibility to both sides of the target so that the charges may be placed flush against the respective target sides.

c. Priming. The simultaneous explosion of both charges is mandatory for optimum results. Crimp nonelectric blasting caps to *equal lengths* of detonating cord. Prime both charges at the *center rear point;* then form a V with the free ends of the detonating cord and attach an electric or nonelectric means of firing.

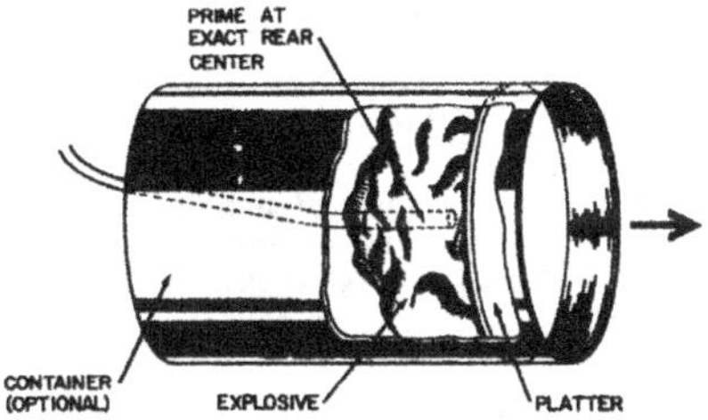

Figure 146. Platter charge.

6. Platter Charge

This device produces the Miznay-Chardin effect. It turns a metal plate into a powerful blunt-nosed projectile (fig. 146). The platter should be steel (preferably round, but square is satisfactory) and should weigh from 2 to 6 pounds.

a. Calculations. Weight of explosive = approximately the weight of the platter.

b. Preparation.

(1) Pack the explosive uniformly behind the platter. A container is not necessary if the explosive can be held firmly against the platter. Tape is acceptable.

(2) Prime the charge from the exact rear center. Cover cap, if any part is exposed, with a small quantity of C4 explosive to insure detonation.

(3) Aim the charge at the direct center of the target.

c. Effect. The effective range (primarily a problem of aim) is approximately 35 yards for a small target. With practice, a demolitionist may hit a 55-gallon drum, a relatively small target, at 25 yards about 90 percent of the time.

7. Grapeshot Charge (Improvised Claymore)

This charge consists of a container, preferably a No. 10 can, projectiles (small pieces of steel), buffer material, an explosive charge, and a blasting cap. These are assembled as shown in figure 147.

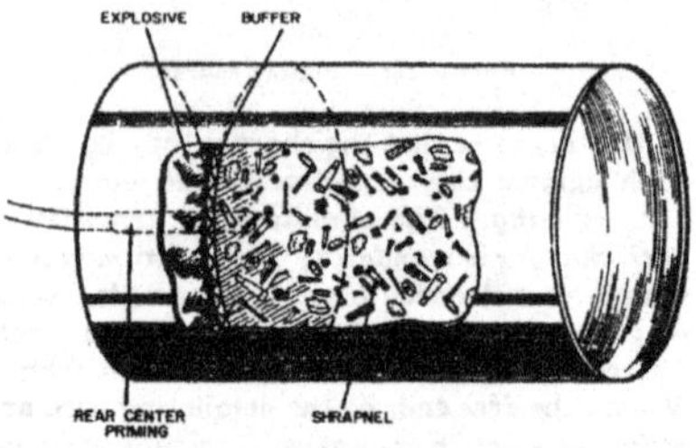

Figure 147. Grapeshot charge.

a. Computation. The weight of the explosive is approximately ¼ × the weight of the projectiles.

b. Preparation.

(1) Assemble the projectiles, a few inches of buffer material—earth, leaves, wood, felt, cloth, cardboard, etc., and the explosive charge. This should be C4, packed firmly.

(2) Prime the charge from the exact rear center. Cover the cap, if any part is exposed, with a small quantity of C4 to insure detonation.

(3) Aim the charge toward the center of the target.

8. Dust Initiator

This device consists of an explosive charge (powdered TNT or C3; C4 will not properly mix with the incendiary), an incendiary mix (2 parts of aluminum powder or magnesium powder to 3 parts of ferric oxide), and a suitable finely-divided organic material (dust) or a volatile fuel such as gasoline called a surround. The dust initiator is most effective in an inclosed space, like a box car or a warehouse or other relatively windowless structure. At detonation, the surround is distributed throughout the air within the target and ignited by the incendiary material.

a. Computation.

(1) Charge size = 1 pound (½ explosive, ½ incendiary mix).

(2) Cover size = 3 to 5 pounds for each 1,000 cubic feet of target. The one-pound charge will effectively detonate up to 40 pounds of cover.

b. Preparation. Powdered TNT may be obtained by crushing it in a canvas bag. The incendiary mix must be thoroughly dispersed throughout the explosive. A great number of dust materials may be used as cover, among which are coal dust, cocoa, bulk powdered coffee, confectioners sugar, tapioca, wheat flour, corn starch, hard rubber dust, aluminum powder, magnesium powder, and powdered soap. If gasoline is used, 3 gallons is the maximum, as more will not disperse evenly in the air and thus give poor results.

9. Improvised Cratering Charge

This charge is a mixture of ammonium nitrate fertilizer containing at least 33 1/3 percent nitrogen and diesel fuel, motor oil, or gasoline at a ratio of 25 pounds of fertilizer to a quart of fuel. The fertilizer must not be damp. From this mixture, improvised charges of almost any size or configuration can be made.

a. Pour the liquid on the fertilizer.

b. Allow the mixture to soak for an hour.

c. Place about half the charge in the borehole. Then place the primer, a primed 1-pound block of TNT, and add the remainder of the charge. Never leave the charge in the borehole for a long period, as accumulated moisture reduces its effectiveness.

d. Detonate the charge.

10. Ammonium Nitrate Satchel Charge

While the cratering charge (para 9 above) is excellent, it is suitable only for cratering. A more manageable charge may be used by mixing ammonium nitrate fertilizer with melted wax instead of oil. The primer is set in place before the mixture hardens.

a. Preparation.

(1) Melt ordinary paraffin and stir in ammonium nitrate pellets, making sure that the paraffin is hot while mixing.

(2) Before the mixture hardens add a half-pound block of TNT or its equivalent as a primer.

(3) Pour the mixture into a container. Shrapnel material may be added to the mixture if desired or attached on the outside of the container to give a shrapnel effect.

b. Use. Because the wax and fertilizer may be molded into almost any size or shape, it may be applied to a great many demolition projects with satisfactory effects.

Section III. COMMODITIES USEFUL FOR MAKING IMPROVISED EXPLOSIVE

11. Introduction

This section deals with materials usable in the manufacture of homemade explosives, incendiaries, and delay devices. Below is a partial list of commodities obtainable commercially that may be used. A glance at this list will show the great complexity of the problem of security forces who must deny these to insurgents. The strictest possible control of their purchase and sale is imperative at the earliest possible moment.

12. Commodity List

The list is as follows:

Ammonium nitrate (fertilizer)
Ammonium perchlorate
Charcoal
Coal
Common match heads
Calcium carbide
Catechol
Dinitrobenzine
"Duco" cement
Flake aluminum
Fuel oil
Glycerin
Hydrogen peroxide (10 volume or higher)
Kerosene
Limed rosin
Liquid floor wax
Lead dioxide
Lead tetraetbyl
Manganese dioxide
Mercury or mercury salts
Nitrobenzine
Nitromethane
Nitrocellulose (pyroxylin)
Sodium chlorate
Sugar
Sawdust
Coal dust
Cocoa
Powdered coffee
Confectioners sugar
Tapioca
Wheat flour
Powdered rice
Cornstarch
Hard rubber dust
Cork dust
Powdered soap
Gasoline
Photoflash powder
Picric acid (certain dye derivatives)
Phenol
Potassium permanganate
Potassium nitrate
Potassium chlorate
Powdered aluminum
Powdered magnesium
Powdered zinc
Paraffin
Petroleum jelly
Pitch
Rosin
Resorcinol
Red phosphorous
Sodium nitrate
Sulfur
Sulfuric acid
Stearic acid
White phosphorous
Nitric acid
Calcium hypochlorite
Turpentine
Potassium dichromate
Sodium peroxide
Nitric acid
Copper sulfate
Carbon disulfide

Plaster of Paris
Ferric oxide
Barium peroxide
Red lead
Ferric sulfate
Aluminum powder
Aluminum sulfate
Naphtha
Silver nitrate powder

Note. As a rule, improvised explosives and incendiaries are more dangerous to handle than conventional explosives. Many mixtures may be ignited or detonated by a single spark, excessive heat, water, or the friction generated by stirring or mixing the ingredients together. Thus, only those who are well informed on the characteristics and reactions of the ingredients should attempt to make improvised explosives.

Section IV. UNDERWATER DEMOLITIONS

13. Introduction

Underwater demolitions involve four basic procedures—reconnaissance in search of obstacles, charge priming, charge placement, and charge initiation.

a. Reconnaissance. As a map may show only the superficial character of a water obstacle, the important information must be obtained by on-site physical reconnaissance by men trained in underwater techniques.

b. Types of Obstacles. Two types of obstacles may be found under water—natural and artificial. Natural obstacles include steep banks, debris, floating logs and brush, underwater ledges, natural craters (particularly at ford sites), rocks, shoals, sandbars, islands, icecrust, and floating ice. Artificial obstacles consist of land mines, boobytraps, floating mines, mines attached to submerged poles, floating obstacles, craters, concrete walls, barbed wire, and conventional concrete and metal obstacles generally found on land but often very effective under water.

14. Priming Underwater Charges

a. Explosive.

(1) *Types.* Tetrytol (M1 chain and M2 blocks), Composition C–3 (M3 and M5 blocks), TNT, and bangalore torpedoes are adaptable to underwater demolitions. Tetrytol may be submerged in water as long as 24 hours without any appreciable effects on its explosive characteristics, while C3 and TNT may be submerged longer if they remain in the original package or are placed in some other sort of container. In addition two U. S. Navy charge assemblies—Mk 133 models 0 and 2 and Mk 135 model 0—are recommended for use under water. They should not be submerged longer than 3 hours before firing, however.

(2) *Mk 133 model 2 demolition charge assembly.*

(a) *Characteristics* (fig. 148).

Container	Color	Size	Weight
Mk 2 model 0 canvas haversack with waterproof, fireproof and mildew-resistant treatment.	Gray	10 x 12 x 9 in	23.5 lb

Explosive: Main charge	Explosive: Booster	Accessories	Remarks
8 blocks H BX–1, 2.5 lb each, strung 1 ft apart on det cord.	50/50 pentolite located at center of block.	10 ft of sash cord with 2 flat hooks for lashing to obstacle; flotation bladder for towing; and tow ring.	Resembles M1 chain demolition charge. Five-foot length of det cord extends from each end of charge.

(b) *Use.* The assembly is a source of eight individual demolition charges—for placing singly or as multiple charges—by cutting the detonating cord chain.

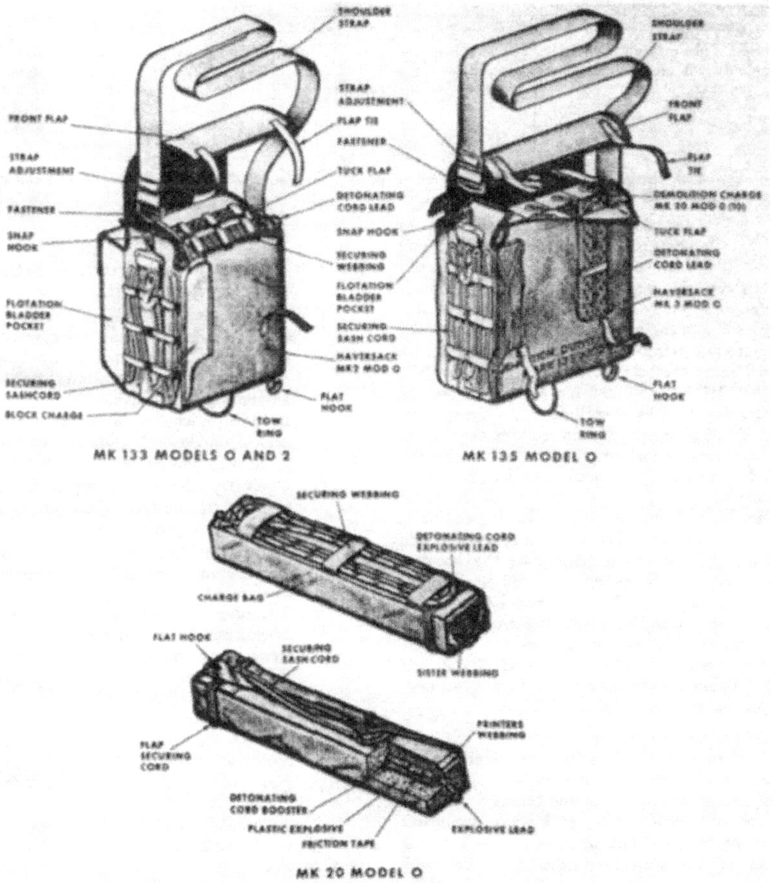

Figure 148. U. S. Navy demolition charge assemblies.

(3) *Mk 135 Model 0 demolition charge assembly.*

(a) *Characteristics* (fig. 148).

(b) *Use.* The individual explosive charge, being tightly wrapped in a canvas bag, cannot be molded into

Container	Color	Size	Weight
Mk 3 model 0 haversack with waterproof, fireproof, and mildew-resistant treatment.	Gray	12 x 14 x 9 in	24 lb (approx)

Explosive		Accessories
Main Charge	Booster (reinforce)	
10 blocks composition C3 (Mk 20 model 0), 2-lb each, in individual canvas charge bags.	5-ft length of reinforced det cord, looped to make a 1-ft booster core, molded into explosive block.	Each charge has a 3.5-ft sash cord and flat hook for lashing to obstacle. Haversack has a 10-ft sash cord and 2 flat hooks for lashing to obstacle, a flotation bladder for towing by swimmer, and a tow ring.

Remarks. Each charge has an 11-ft length of reinforced detonating cord—5 ft is looped and molded in the block and the remaining 6 ft is an explosive lead.

any desired shape, but it can be bent or curved into close contact with the surface of the target.

b. Priming. Because detonating cord, though watersoaked, may be detonated if initiated at a dry end, it is the most satisfactory of all firing systems for priming explosive charges used on underwater obstacles. M1 chain, M2, M3, M5, M5A1, M112, and TNT blocks are primed by detonating cord as described in paragraphs 64 and 65. Bangalore torpedo sections are primed as described in paragraph 69*b*. Although the M1 chain demolition block may be primed by means of a branch line attached to the detonating cord chain, priming is more positive if the detonating cord primer is wrapped and tied around the end of the block over the booster. The same is true of priming the M2 block. The 2.5-lb. block of the Mk 133 model 0 assembly should be primed by wrapping and tying the detonating cord around the center over the booster; while the M3, M5, M5A1, M112, and TNT, having no boosters, may be primed anywhere on the block (para 64).

15. Charge Placement

a. In underwater demolitions both single and multiple charges are used, depending on the size and configuration of the target. The size of the charge is computed by the applicable table or formula, but because of the tamping effect of the water, charges on underwater targets require only about 1/3 as much explosive as untamped charges used on similar targets on land. Water-tamped charges, however, require the same amount of explosive as tamped charges on land.

b. Multiple charges are connected by detonating cord branch lines attached by means of *knots* to a main line or a ring main as described in paragraph 70*b* through *d*.

c. The following procedures should be followed, if possible, in underwater demolitions. Charges should be placed so that—

(1) Their pressure waves will not counteract each other. To avoid this, charges are placed in staggered lines—the adjacent charges containing different amounts of explosive. For example, if the first charge is 20 pounds, the second should be 10 pounds, the third, 20 pounds and so on.

(2) Their fragments will not be thrown toward friendly troops and equipment.

(3) They will not throw heavy debris in or partially obstruct a path to be taken by friendly troops or equipment. For example, a charge placed on top of a boulder may merely leave the large fragments that remain as obstructions, and a charge placed directly under a boulder may form a few large fragments and a large crater underneath, thus enlarging instead of removing the obstacle. On the other hand, a snakehole charge properly placed may move the boulder and fragments from the path. These principles also apply to the removal of other obstacles.

16. Charge Initiation

Detonating cord systems used underwater may be initiated by an electric or nonelectric detonating assembly attached to a dry end of the ring main by means of a square knot (para 70*a*, fig. 85) or by means of an electric or nonelectric firing system attached directly to the ring main.

INDEX

By Order of the Secretary of the Army:

HAROLD K. JOHNSON,
General, United States Army,
Chief of Staff.

Official:

KENNETH G. WICKHAM,
Major General, United States Army,
The Adjutant General.

Distribution:

To be distributed in accordance with DA Form 12-11 for Explosives and Demolition.

☆ U.S. GOVERNMENT PRINTING OFFICE: 1967—/7258A

Made in United States
North Haven, CT
10 November 2026

22095627R10104